Dr Catherine Dawson has been a researcher and writer since the mid-1980s and has taught research methods courses at university to undergraduate students, postgraduate students and to community groups. She has also taught a course called 'Reflecting on your Enquiry Skills', which was aimed at members of the public wishing to carry out a research project but who had no formal education or training. Dr Dawson has also written extensively for academic journals on a wide range of subjects including research methodology. In this book she explains complicated issues in a way that can be understood by anyone who is interested in their topic.

Introduction to Research Methods

5th Edition

....

Dr Catherine Dawson

ROBINSON

First published in Great Britain in 2002 by How To Books Ltd

This revised and updated edition published in Great Britain in 2019 by Robinson

9 8 7 6 5

A CIP catalogue record for this book is available from the British Library

ISBN: 978-1-40871-105-7

Typeset in Sentinel and Scala sans by Mousemat Design Limited

Printed and bound in Great Britain by Clays Ltd, Elcograf S.p.A.

Papers used by Robinson are from well-managed forests and other responsible sources

Robinson
An imprint of
Little, Brown Book Group
Carmelite House
50 Victoria Embankment
London EC4Y 0DZ

An Hachette UK Company
www.hachette.co.uk

www.littlebrown.co.uk

How To Books are published by Robinson, an imprint of Little, Brown Book Group. We welcome proposals from authors who have first-hand experience of their subjects. Please set out the aims of your book, its target market and its suggested contents in an email to howtobooks@littlebrown.co.uk

Contents

List of Illustrations

Tables

Figures

Preface

Welcome to the fifth edition of this book, which has been updated fully to include more useful tips and advice, new references and updated contact details of relevant organisations. It also includes a new chapter on conducting experiments in social research, which will be of particular relevance to students and researchers from health disciplines, psychology and economics, for example. This book is a practical, down-to-earth guide for people who wish to conduct social research. It is aimed at those new to research and assumes no prior knowledge of the issues covered. It will also appeal to those people who have already conducted some research and who are interested in finding out more about other research methods that are available to them.

For the purpose of this book, social research is defined as the deliberate study of other people for the purposes of increasing understanding and/or adding to knowledge.

This deliberate study could cover many different areas. As a researcher, you might be interested in attitudes and behaviour – why do people think in a certain way and why do they behave in a certain way? Or you might be interested in numbers – how many people use a service? Perhaps you need to try to predict how this number of people could be increased so that you can obtain funding for your service. Or you might be fascinated by the personal history of a neighbour and have a burning desire to record her history and pass it on to others.

We all have different reasons for conducting research. Some of us might have to undertake a project as part of our course work. Others might have to conduct a study as part of our employment. Some of us may be fascinated by something we've observed and want to find out more. This book offers advice on how to turn your ideas into a workable project and how to keep motivation levels high, especially if you have no real inclination to become a social researcher. It discusses the issues involved in thinking about your research and defining your project, before moving on to the methods – how do you actually *do* your research, analyse your findings and report the results?

Over the decades there has been a great deal of discussion on what constitutes social research, how it should be conducted and whether certain methods are 'better' than others. Although I have touched on some of these issues in the relevant chapters, it is not possible or desirable to go into any greater detail in this book. Therefore, I have included further reading sections at the end of the relevant chapters for those of you who wish to follow up these issues.

I have been a researcher since undertaking an MA in Social Research in 1987. Working within both further and higher education and as a freelance researcher, I have been involved in a variety of projects in the areas of education, housing and community research. I have taught research methods to adults returning to education and conducted in-house training for employees who need to carry out their own research.

Becoming a successful researcher is a continual learning process in which we all make mistakes. So don't worry if your first project doesn't run as smoothly as you might wish. Instead, remember that undertaking a research project can be fascinating, rewarding and exciting. I hope that you enjoy it as much as I have done and I wish you every success in your project.

Dr Catherine Dawson

1

How to Define Your Project

Before you start to think about your research, you need to ask yourself a few questions.

Asking questions

Why have I decided to do some research?

If the answer to this question is because you have been told to do so, either by your tutor or by your boss, you need to think about how you're to remain motivated throughout your project. Research can be a long process and take up much of your time. It is important to stay interested in what you're doing if you are to complete your project successfully. However, if you want to conduct some research because something has fascinated you, or you have identified a gap in the research literature, then you are lucky and should not have a problem with motivation.

How can I remain interested in my research?

The obvious answer to this is to choose a topic which interests you. Most of you do have this choice within the limitations of your subject – be creative and think about something that will fascinate you. However, if you have had the topic chosen for you, try to choose a research *method* that interests you.

How do I choose a research method?

As you go on to read this book you will become more familiar with the different methods and should be able to find something in which you are interested.

The following questions will help you to start to think about these issues:

- Did you enjoy mathematics at school? If so, perhaps you might be interested in delving deeper into statistical software or other types of data analysis?

- Have you ever taken part in a focus group or been interviewed by a market researcher? Would you find it interesting to conduct your own focus groups or interviews?

- Have you been fascinated by a particular group of people? Would you like to immerse yourself in their culture and learn more?

- Do you enjoy filling in questionnaires? Would you like to design your own questionnaire and perhaps conduct a postal or online survey?

- Do you enjoy undertaking experiments, following set rules and procedures to collect data in a systematic and focused way?

What personal characteristics do I have that might help me to complete my research?

Think about your personal characteristics, likes and dislikes, strengths and weaknesses when you're planning your research.

The following questions will help you to do this:

- Are you good with people?

- Do you prefer written communication or face-to-face interaction?

- Do you love or loathe mathematics and statistics?

- Do people feel at ease with you and are they willing to confide in you?

- Do you like to number crunch?

- Do you like to conduct online research?

What skills and experience do I have that might help in my research?

If your research is to be employment based, the chances are you will have work experience that you'll find useful when conducting your research project. This is valid experience and you should make the most of it when planning your research.

Even if your project is not employment based, all of you will have other skills and experience that will help. For example, if you have been a student for three years, you will have developed good literature search skills that will be very useful in the research process. Some of you may have developed committee skills, organisation skills and time management expertise. All of these will be extremely useful in your research.

It is important to think about your existing skills in relation to your proposed project, as it will help you to think about whether your knowledge, experience and skills will help you to address the problem you have identified.

Thinking about your project

Many research projects fail because people don't take enough time to think about the issues involved before rushing to start the work. It is extremely important to spend time *thinking* about your project before you move on to the planning stage. Through careful thought you should stop yourself wasting time and energy on inappropriate methods as your research progresses. Consider the following example:

EXAMPLE 1: JAMES

James wanted to find out about students' experiences of housing in his university town. He designed and sent out a questionnaire to 1,000 students. When the replies started to come in, he realised that the questionnaires weren't generating the type of information in which he was interested. When he talked through his concerns with his tutor, it emerged that James was really interested in attitudes towards, and experiences of, rented accommodation. Instead, he was only finding out about how many students lived in private rented accommodation and whether they had had 'good' or 'bad' experiences. The questionnaire left him unable to delve deeper into what these experiences were, how students coped with them and how these experiences affected their attitude towards private rented accommodation. His questionnaire had been poorly designed and was not generating this type of information.

James had to scrap the questionnaire and construct another that he combined with a number of one-to-one interviews to get more in-depth information. He had spent three months designing and administering a questionnaire that had not produced the type of information he required. If he had spent more time thinking about the research, especially coming to terms with the difference between *qualitative* and *quantitative* research, he would have saved himself a lot of time and energy (see Chapter 2).

Understanding the five 'Ws'

When you start to think about your research project, a useful way of remembering the important questions to ask is to think of the five 'Ws':

- What?
- Why?
- Who?
- Where?
- When?

Once you have thought about these five 'Ws' you can move on to think about *how* you are going to collect your data.

What?

What is your research? This question needs to be answered as specifically as possible. One of the hardest parts in the early stages is to be able to define your project – so much research fails because the researcher has been unable to do this.

TIP

Sum up, in one sentence only, your research. If you are unable to do this, the chances are your research topic is too broad, ill thought out or too obscure.

Why?

Why do you want to do the research? What is its purpose? Okay, you might have been told to do some research by your tutor or by your boss, but there should be another reason why you have chosen your particular subject. Reasons could include the following:

- You are interested in the topic.
- You have identified a gap in the literature.
- You want to obtain funding for a particular service or enterprise and you need to find out whether there is a demand for what you are proposing.
- You need to conduct some research to aid decision making.

Whatever your reason, think very carefully about why you are doing the research, as this will affect your topic, the way you conduct the research and the way in which you report the results. You should consider the following points:

- If you're conducting the research for a university dissertation or project, does your proposed research provide the opportunity to reach the required intellectual standard? Will your research generate enough material to write a dissertation of the required length? Will your research generate too much data that would be impossible to summarise into a report of the required length?
- If you're conducting research for funding purposes, have you found out whether your proposed funding body requires the information to be presented in a specific format? If so, you need to plan your research in a way that will meet that format.

Speak to as many people as possible about your research, including tutors, fellow students, colleagues or friends. Tell them why you have chosen the project and ask them for their thoughts. This will help you to reflect upon, and develop, your own ideas.

Who?

Who will be your participants? (In this book, people who take part in research will be called **participants** or **respondents**, but you can use 'subjects' if you prefer.)

At this stage of the research process, you needn't worry too much about exactly how many participants will take part in your research, as this will be covered later (see Chapter 5). However, you should think about the type of people with whom you will need to get in touch and whether it will be possible for you to contact them. If you have to conduct your research within a particular time scale, there's little point choosing a topic that would include people who are difficult or expensive to contact. Also, bear in mind that online research provides opportunities for contacting people cheaply, especially if you're a student with free internet access.

Where?

Where are you going to conduct your research? Thinking about this question in geographical terms will help you to narrow down your research topic. Also, you need to think about the resources in terms of available budget and time. If you're a student who will not receive travel expenses or any other out-of-pocket expenses, choose a location close to home, college or university. If you're a member of a community group on a limited budget, only work in areas within walking distance, as this will cut down on travel expenses.

Also, you need to think about the venue. If you're going to conduct interviews or focus groups, where will you hold them? Is there a room at your institution that would be free of charge, or are you going to conduct them in participants' own homes? Would it be safe for you to do so? Would you be comfortable doing so?

If you've answered 'no' to either of these last two questions, maybe you need to think again about your research topic. In 20 years I have encountered only one uncomfortable situation in a stranger's home. It can happen and you must never put yourself in a dangerous situation. Think very carefully about whether your chosen topic and method might have an influence on personal safety.

When?

When are you going to do your research? Thinking about this question will help you to sort out whether the research project you have proposed is possible within your time scale. It will also help you to think more about your participants, when you need to contact them and whether they will be available at that time. For example, if you want to go into schools and observe classroom practice, you wouldn't choose to do this research during the summer holiday. It might sound obvious, but I have found some students present a well-written research proposal which, in practical terms, will not work because the participants will be unavailable during the proposed data collection stage.

Summarising your research

Once you have thought about these five 'Ws', try to sum up your proposed project in one sentence. When you have done this, take it to several people, including your boss and/or tutor, and ask them if it makes sense. Do they understand what your research is about? If they don't, ask them to explain their confusion, then revise your statement and take it back to them.

I can't overemphasise the importance of this stage of the research process. If you get it right now, you will find that the rest of your work should flow smoothly. However, if you get it wrong, your problems could well escalate. The following exercise will help you to think more about these issues.

EXERCISE 1:

Have a look at the three projects below and see if you can spot any potential problems. What questions would you ask to make the researchers focus in on their proposed project? Do you have any suggestions for the improvement of these statements?

Statement 1: *This research aims to find out what people think about television.*

Statement 2: *My project is to do some research into Alzheimer's disease, to find out what people do when their relatives have it and what support they can get and how nurses deal with it.*

Statement 3: *We want to find out how many of the local residents are interested in a play scheme for children during the summer holiday.*

Points to consider

Statement 1: *This research aims to find out what people think about television.* This proposed project is both broad and obscure. My first two questions would be: what people and what television? Then I would ask: what is the purpose of this research? Who would be interested in the results? TV companies already employ market researchers to conduct a great deal of research into public viewing, and they have much larger budgets available to them. There's little point in repeating research if it cannot be improved upon.

However, if the researcher has an interest in this particular issue, or is perhaps on a media studies course, there are a number of ways in which this research could become more manageable. For example, the research could focus on a particular type of programme and/or a particular type of person, as the following examples suggest:

- She could decide to show an Open University (OU) programme to potential OU students and find out what they thought about the programme in a series of focus groups.
- She could choose children's programming and find out what teachers think about the educational value of these programmes.
- She could ask business people what they think about a programme aimed specifically at the business community.
- She could ask fellow students to keep a diary of their television viewing over a week and then interview them about their viewing habits.

There are many different possibilities within this field. The researcher needs to decide exactly where her interests lie and focus in on those interests.

Statement 2: *My project is to do some research into Alzheimer's disease, to find out what people do when their relatives have it and what support they can get and how nurses deal with it.*

The main problem with this statement is the grammar. The topic itself is more focused, as the researcher has mentioned, specifically, the areas he wishes to consider – nurses' attitudes, carers' experiences and available support. His topic is immediately more manageable because he is only considering nurses or carers who come into contact with sufferers of Alzheimer's disease. However, he needs to think about whether he is going to consider hospitals, residential homes, or both, and in what areas. Also, is he going to contact people who look after their relatives at home?

Although, on the surface, this project appears more manageable, this researcher has a major point to consider. In the UK all social research that is carried out on health care premises comes under the jurisdiction of Research Ethics Committees. These committees were set up to ensure that research does not harm patients in any way and that it is done in their best interests. In the USA a similar function is carried out by Institutional Review Boards. (See useful websites at the end of this chapter.)

This means that the researcher would have to get his project approved by the appropriate committee before he could go ahead with the research, and it is not guaranteed that his project would be given approval. As he would have to submit a full and detailed proposal to the committee, he could be conducting a lot of preliminary work, only to be turned down. You would need to think carefully whether this is a route you wish to take, and if so, you would need to obtain the appropriate advice before committing yourself.

TIP

The Medical Research Council has produced a Health Research Authority Decision Tool *(www.hra-decisiontools.org.uk/research)* that will help you to find out whether your research will need Ethics Committee approval in the UK.

Statement 3: *We want to find out how many of the local residents are interested in a play scheme for children during the summer holiday.*

This project put forward by a tenants' association appears to be straightforward and manageable, although there are still several issues that need addressing. My first question for this topic would be: do you really want to find out how many of the local residents are interested, or do you want to find out the interests of *residents with children of the appropriate age who would actually use the scheme?* If the latter is the case, this narrows down the research population and makes it more manageable.

Finding out whether someone is *interested* in something is not actually the same as finding out whether someone would *use* the service. For example, I might think a play scheme is a good idea for other children as it might keep them off the streets, but not for my little darlings, who are too occupied with their computer. If I said 'yes, I am interested', this could be misleading as I have no intention of using the service. However, if the purpose of the research is to obtain funding for the scheme, then the more people who express an interest, the better, although the tenants' association would have to be careful not to produce misleading information.

I would also find out whether the tenants' association was interested only in the issue of how many people were interested in it and would use the play scheme. If they were doing this research anyway, would it be a valuable addition to find out what sort of scheme residents would like, and what activities their children would like? Would residents have any reservations about sending their children? If they do have reservations, what are they? Who would residents want to run the scheme? Would they be willing to provide help and support themselves?

Summary

- You must take time to think about your research, as this will save you problems later.

- When you're thinking about your research, ask yourself the five 'Ws':
  - *What* is my research?
  - *Why* do I want to do the research?
  - *Who* are my research participants?
  - *Where* am I going to do the research?
  - *When* am I going to do the research?

- Sum up your research project in one sentence.

- Discuss your sentence with your tutor or boss and revise it if there is any confusion.

Useful websites

www.hra.nhs.uk/news/rec

The Health Research Authority provides detailed information about applying for approval and the review process for medical research in the UK. Contact details for all NHS Research Ethics Committees can be obtained from this website.

www.fda.gov

This is the website of the US Food and Drug Administration. From this site you can obtain more information and guidance about Institutional Review Boards in the US and find out about conducting biomedical research with human participants.

2
How to Decide Upon a Methodology

Once you have answered the five 'Ws' you can go on to think about *how* you're going to do your research. The first thing you need to do is to think about your **research methodology**. This is the philosophy or the general principle that will guide your research. It is the overall approach to studying your topic and includes issues you need to think about such as the constraints, dilemmas and ethical choices within your research.

Now that you have read Chapter 1, some of these issues will be fresh in your mind. Your research **methodology** is different to your research **methods** – these are the tools you use to gather data, such as questionnaires or interviews, and these will be discussed in Chapter 3.

Recognising qualitative and quantitative research

When you start to think about your research methodology, you need to think about the differences between **qualitative** and **quantitative** research.

Qualitative research explores attitudes, behaviour and experiences through such methods as interviews or focus groups. It attempts to get an in-depth opinion from participants. As it is attitudes, behaviour and experiences that are important, fewer people take part in the research, but the contact with these people tends to last a lot longer.

There are many different methodologies that come under the umbrella term of

> **TIP**
>
> Try to become familiar with the different types of methodology. When particular research results are reported in newspapers or on television, think about what you are being told. Can you work out whether qualitative or quantitative methodologies have been used to inform the research?

qualitative research. Examples of some of these methodologies are summarised below. If you wish to pursue any of these in more depth, useful references are included at the end of this chapter.

Quantitative research generates statistics through the use of large-scale survey research, using methods such as questionnaires or structured interviews. If a market researcher has stopped you on the streets, or you have filled in a questionnaire that has arrived through the post, this falls under the umbrella term of quantitative research. This type of research reaches many more people, but the contact with those people is much quicker than it is in qualitative research.

Understanding the methodological debate

Over the years there has been a large amount of complex discussion and argument surrounding the topic of social research methodology and the theory of how inquiry should proceed. Much of this debate has centred on the issue of qualitative versus quantitative inquiry – which might be the best and which is more 'scientific'.

Different methodologies become popular at different social, political, historical and cultural times in our development, and, in my opinion, all methodologies have their specific strengths and weaknesses. These should be acknowledged and addressed by the researcher.

At the end of this chapter, references are given if you are interested in following up any of these issues. Certainly, if you were to do so, it would help you to think about your research methodology in considerable depth.

Choosing a methodology

Don't fall into the trap that many beginning (and experienced) researchers do of thinking that quantitative research is 'better' than qualitative research. Neither is better than the other – they are just different and both have their strengths and weaknesses. Both also depend on the skills, training and experiences of the researcher.

What you will find, however, is that your instincts probably lean you towards one rather than the other. Listen to these instincts, as you will find it more productive to conduct the type of research with which you will feel comfortable, especially if you're to keep your motivation levels high. Also, be aware of the fact that your tutor or boss might prefer one type of research over the other. You might have a harder time justifying your chosen methodology if it goes against their preferences.

TIP

Discuss your methodological thoughts with your tutor or boss. Listen to their opinions and gauge their response. If you decide to go ahead with a methodology about which they are unaware, or with which they disagree, be prepared to fight your cause. This will involve careful thought and consideration.

EXAMPLES OF QUALITATIVE RESEARCH METHODOLOGIES

Action research

Some researchers believe that action research is a research method, but in my opinion it is better understood as a methodology. In action research, the researcher works in close collaboration with a group of people to improve a situation in a particular setting. The researcher does not 'do' research 'on' people, but instead works with them, acting as a facilitator. Therefore, good group management skills and an understanding of group dynamics are important skills for the researcher to acquire. This type of research is popular in areas such as organisational management, community development, education and agriculture.

Action research begins with a process of communication and agreement between people who want to change something together. Obviously, not all people within an organisation will be willing to become co-researchers, so action research tends to take place with a small group of dedicated people who are open to new ideas and willing to step back and reflect on these ideas. The group then moves through four stages of planning, acting, observing and reflecting. This process may happen several times before everyone is happy that the changes have been implemented in the best possible way.

In action research, various types of research method may be used, for example: the diagnosing and evaluating stage may use questionnaires and interviews, whereas focus groups may be used to gauge opinion on the proposed changes.

Ethnography

Ethnography has its roots in anthropology and was a popular form of inquiry at the turn of the century when anthropologists travelled the world in search of remote tribes. The emphasis in ethnography is on describing and interpreting cultural behaviour.

Ethnographers immerse themselves in the lives and culture of the group being studied, often living with that group for months on end. These researchers participate in a group's activities while observing its behaviour, taking notes, conducting interviews, analysing, reflecting and writing reports – this may be called **fieldwork** or **participant observation**. Ethnographers highlight the importance of the written text because this is how they portray the culture they are studying.

Feminist research

There is some argument about whether feminist inquiry should be considered a methodology or an epistemology, but in my opinion it can be both. (As we have seen, *methodology* is the philosophy or the general principle that will guide your research. **Epistemology**, on the other hand, is the study of the nature of knowledge and justification. It looks at where knowledge has come from and how we know what we know.)

Feminist researchers argue that for too long the lives and experiences of women have been ignored or misrepresented. Often, in the past, research was conducted on male 'subjects' and the results generalised to the whole population. Feminist researchers critique both the research topics and the methods used; especially those that emphasise objective, scientific 'truth'.

With its emphasis on participative, qualitative inquiry, feminist research has provided a valuable alternative framework for researchers who have felt uncomfortable with treating people as research 'objects'. Under the umbrella of feminist research are various different standpoints – these are discussed in considerable depth in some of the texts listed at the end of this chapter.

Grounded theory

Grounded theory is a methodology that was first laid out in 1967 by two researchers named Glaser and Strauss. It tends to be a popular form of inquiry in the areas of education and health research. The emphasis in this methodology is on the generation of theory which is grounded in the data – this means that it has emerged from the data. This is different from other types of research which might seek to test a hypothesis that has been formulated by the researcher. (See Chapter 11.)

In grounded theory, methods such as focus groups and interviews tend to be the preferred data collection method, along with a comprehensive literature review that takes place throughout the data collection process. This literature review helps to explain emerging results.

In grounded theory studies, the number of people to be interviewed is not specified at the beginning of the research. This is because the researcher, at the outset, is unsure of where the research will take her. Instead, she continues with the data collection until 'saturation' point is reached, that is, no new information is being provided. Grounded theory is therefore flexible and enables new issues to emerge that the researcher may not have thought about previously.

Matching methodology with topic

So, how do you decide which is the best methodology for your research? Perhaps the easiest way to do this is to decide first of all whether you should consider qualitative or quantitative research. Have another look at the five 'Ws' discussed in Chapter 1. If you have not already done so, go through each question in relation to your own research. Once you have done this, clues will start to emerge about what is the best form of inquiry for you.

First of all, have a look at the words you have used. Certain words help to suggest a leaning towards qualitative research, others towards quantitative research. For example, if you have written 'how many', 'test', 'verify', 'how often' or 'how satisfied', this suggests a leaning towards quantitative research. If you have written words such as 'discover', 'motivation', 'experiences', 'think/thoughts', 'problems' or 'behave/behaviour', this suggests a leaning towards qualitative research.

However, you may find that you have written a combination of these words, which could mean two things. Firstly, you might want to think about combining both qualitative and quantitative research, which is called **triangulation** or **mixed methods research**. Many researchers believe this is a good way of approaching research, as it enables you to counteract the weaknesses in both qualitative and quantitative research. Secondly, it could mean that your ideas are still unclear and that you need to focus a little more.

To help you understand the thought processes involved in these decisions, let's return to the exercise given in the previous chapter:

EXAMPLE 2: REVISED STATEMENTS

Original statement 1: *This research aims to find out what people think about television.*
After having thought about how to focus her topic, make the project more manageable and produce a worthwhile piece of research, the researcher came up with the following revised statement:

Revised statement 1: *This research aims to find out what primary school teachers think about the educational value of the* Teletubbies' *television programme.*
This research topic is now well focused. When the student suggested this research, it was also very topical – the *Teletubbies* had been released only four weeks prior to the research and complaints about their language were filling the national media. The main clue to the methodology is the word 'think'. The student wishes to get an in-depth opinion, but is not concerned with speaking to a large number of primary school teachers. This suggests a qualitative form of inquiry.

Original statement 2: *My project is to do some research into Alzheimer's disease, to find out what people do when their relatives have it and what support they can get and how nurses deal with it.*

This researcher decided to narrow down his topic. Also, he found out whether he needed to obtain Health Research Authority Approval by visiting *www.hra.nhs.uk/research-community/before-you-apply*. This website gives advice and guidance about whether ethical approval is required for a research project in the UK and, if so, provides information about the application process.

Revised statement 2: *The aim of this research is to find out how many relatives of Alzheimer's patients use the Maple Day Centre, and to ascertain whether the service is meeting their needs.*
Again this topic is now much better focused. The research population is limited to relatives of Alzheimer's patients who use the Maple Day Centre. One clue to the methodology is in the words 'how many', which suggests a quantitative study. However, the researcher is also interested in finding out whether the service meets their needs, which requires some more in-depth inquiry. This suggests a combination of qualitative and quantitative inquiry (mixed methods research).

Original statement 3: *We want to find out how many of the local residents are interested in a play scheme for children during the summer holiday.*
The tenants' association thought carefully about the issues in which they were interested, eventually coming up with the following revised statement:

Revised statement 3: *This research aims to find out how many people from our estate are interested in, and would use, a children's play scheme in the school summer holiday.*
Again, the clue in this example is 'how many'. The tenants' association wanted to obtain funding for their play scheme and felt that it was important to gather statistics that they could take to possible funding organisations. This suggests a quantitative study.

Summary

- The research *methodology* is the philosophy or general principle that guides the research.
- Research *methods* are the tools you use to gather your data.
- *Qualitative research* explores attitudes, behaviour and experiences.
- Examples of qualitative methodologies include action research, ethnography, feminist research and grounded theory.
- *Quantitative research* generates statistics through the use of large-scale survey research.
- Neither qualitative nor quantitative research is better – they are just different. Both have their strengths and weaknesses.
- Your own intuition and the words you use will give pointers to whether qualitative or quantitative research is more appropriate for your chosen project.
- The terms '*triangulation*' or 'mixed methods research' are used when a combination of qualitative and quantitative forms of inquiry are used.

Further reading

The theoretical and philosophical issues raised in this chapter are detailed and complex and cannot be discussed in depth in this book. However, if you wish to pursue any of these topics, some of the useful publications are listed below under the relevant topics.

Research methodologies

Clough, P. and Nutbrown, C. (2012) *A Student's Guide to Methodology*, 3rd edition, London: Sage.
Creswell, J.W. (2013) *Research Design: Qualitative, Quantitative, and Mixed Methods Approaches*, 4th edition, Thousand Oaks, CA: Sage.

Qualitative research

Over recent years there has been a great deal of innovation in the use of qualitative methodologies. Listed below are some of the more traditional texts and a selection of the newer, innovative texts.

Denzin, N.K. and Lincoln, Y.S. (eds) (2017) *The Sage Handbook of Qualitative Research*, 5th edition, Thousand Oaks, CA: Sage.

Higgs, J., Armstrong, H. and Horsfall, D. (2001) *Critical Moments in Qualitative Research*, Oxford: Butterworth-Heinemann.

Hollway, W. and Jefferson, T. (2012) *Doing Qualitative Research Differently: A Psychosocial Approach*, 2nd edition, London: Sage.

Schwandt, T. (2015) *The Sage Dictionary of Qualitative Inquiry*, 4th edition, Thousand Oaks, CA: Sage.

Quantitative research

De Vaus, D.A. (2014) *Surveys in Social Research*, 6th edition, Abingdon, Oxon: Routledge.

Fowler, F. (2013) *Survey Research Methods*, 5th edition, Thousand Oaks, CA: Sage.

Lance, C.E. and Vandenberg, R.J. (eds) (2015) *More Statistical and Methodological Myths and Urban Legends*, New York: Routledge Academic.

Sapsford, R. (2006) *Survey Research*, 2nd edition, London: Sage.

Action research

Dadds, M. and Hart, S. (eds) (2001) *Doing Practitioner Research Differently*, London: Routledge Falmer.

McNiff, J. (2000) *Action Research in Organisations*, London: Routledge.

Reason, P. and Bradbury, H. (2013) *The Sage Handbook of Action Research: Participative Inquiry and Practice*, 2nd edition, London: Sage.

Schmuch, R. (ed) (2008) *Practical Action Research: A Collection of Articles*, 2nd edition, Thousand Oaks, CA: Corwin Press.

Ethnography

Atkinson, P. (2015) *For Ethnography*, London: Sage.

Davies, C.A. (2007) *Reflexive Ethnography: A Guide to Researching Selves and Others*, 2nd edition, London: Routledge.

Fetterman, D. (2010) *Ethnography: Step-by-Step*, 3rd edition, Thousand Oaks, CA: Sage.

Hammersley, M. and Atkinson, P. (2007) *Ethnography: Principles in Practice*, 3rd edition, London: Routledge.

Wolcott, H.F. (2008) *Ethnography: A Way of Seeing*, 2nd edition, Walnut Creek, CA: Altamira.

Feminist research

Harding, S. and Hintikka, M. (eds) (2003) *Discovering Reality: Feminist Perspectives on Epistemology, Metaphysics, Methodology, and Philosophy of Science*, Dordrecht: Kluwer Academic Publishers.

Kleinman, S. (2007) *Feminist Fieldwork Analysis*, Thousand Oaks, CA: Sage.

Letherby, G. (2003) *Feminist Research in Theory and Practice*, Buckingham: Open University Press.

Stanley, L. and Wise, S. (1993) *Breaking Out Again: Feminist Ontology and Epistemology*, London: Routledge.

Grounded theory

Birks, M. and Mills, J. (2015) *Grounded Theory*, London: Sage.

Corbin, J. and Strauss, A. (2015) *Basics of Qualitative Research: Techniques and Procedures for Developing Grounded Theory*, 4th edition, Thousand Oaks, CA: Sage.

Glaser, B. and Strauss, A. (1999) *Discovery of Grounded Theory: Strategies for Qualitative Research*, Chicago: Aldine Transactions.

3
How to Choose Your Research Methods

As we have seen in the previous chapter, **research methods** are the tools you use to collect your data. Before you decide which would be the most appropriate methods for your research, you need to find out a little more about these tools. This chapter gives a description of the methods of interviewing, focus groups, questionnaires, participant observation and experiments. Chapters 7–11 will go on to describe in detail how to use each of these methods.

Using interviews

In social research there are many types of interview. The most common of these are **unstructured**, **semi-structured** and **structured** interviews. If you want to find out about other types of interview, relevant references are given at the end of this chapter.

Unstructured interviews

Unstructured or **in-depth** interviews are sometimes called **life history** interviews. This is because they are the favoured approach for life history researchers. In this type of interview, the researcher attempts to achieve a holistic understanding of the interviewees' point of view or situation. For example, if you want to find out about a Polish man's experiences of a concentration camp during the war, you're delving into his life history. Because you are unsure of what has happened in his life, you need to ask as few questions as possible, so as to enable him to talk freely. It is for this reason that this type of interview is called unstructured – the participant is free to talk about what he or she deems important, with little directional influence from the researcher. This type of interview can only be used for qualitative research.

As the researcher tries to ask as few questions as possible, people often assume that this type of interviewing is the easiest. However, this is not necessarily the case. Researchers have to be able to establish **rapport** with the participant – they have to be trusted if someone is to reveal intimate life information. This can be difficult and takes tact, diplomacy and perseverance. Also, some people find it very difficult to remain quiet while another person talks, sometimes for hours on end.

In unstructured interviews, researchers need to remain alert, recognising important information and probing for more detail. They need to know how to steer someone tactfully back from totally irrelevant digressions. Also, it is important to realise that unstructured interviewing can produce a great deal of data that can be difficult to analyse.

Semi-structured interviews

Semi-structured interviewing is perhaps the most common type of interview used in qualitative social research. In this type of interview, the researcher wants to know specific information that can be compared and contrasted with information gained in other interviews. To do this, the same questions need to be asked in each interview. However, the researcher also wants the interview to remain flexible so that other important information can still arise.

For this type of interview, the researcher produces an **interview schedule** (see Chapter 7). This may be a list of specific questions or a list of topics to be discussed. This is taken to each interview to ensure continuity. In some research, such as a grounded theory study, the schedule is updated and revised after each interview to include more topics that have arisen as a result of the previous interview. (See Chapter 2.)

Structured interviews

Structured interviews are used frequently in market research. Have you ever been stopped in the street and asked about washing powder or which magazines you read? Or have you been invited into a hall to taste cider or smell washing-up liquid? The interviewer asks you a series of questions and ticks boxes according to your response. This research method is highly structured – hence the name. Structured interviews are used in quantitative research and can be conducted face-to-face, online or over the telephone, sometimes with the aid of laptops or mobile devices.

Conducting focus groups

Focus groups may be called **discussion groups** or **group interviews**. A number of people are asked to come together in a group to discuss a certain issue. For example, in market research this could be a discussion centred on new packaging for a breakfast cereal, in social research this could be to discuss adults' experiences of school, or in political research this could be to find out what people think about a particular political leader.

The discussion is led by a **moderator** or **facilitator** who introduces the topic, asks specific questions, controls digressions and stops break-away conversations. She makes sure that no one person dominates the discussion, while trying to ensure that each of the participants makes a contribution. Focus groups may be recorded using visual or audio recording equipment.

TABLE 1:
THE FOCUS GROUP METHOD: ADVANTAGES AND DISADVANTAGES

ADVANTAGES	DISADVANTAGES
Can receive a wide range of responses during one meeting.	Some people may be uncomfortable in a group setting and nervous about speaking in front of others.
Participants can ask questions of each other, lessening impact of researcher bias.	Not everyone may contribute.
Helps people to remember issues they might otherwise have forgotten.	Other people may contaminate an individual's views.
Helps participants to overcome inhibitions, especially if they know other people in the group.	Some researchers may find it difficult or intimidating to moderate a focus group.
The group effect is a useful resource in data analysis.	Venues and equipment can be expensive.
Participant interaction is useful to analyse.	Difficult to extract individual views during the analysis.

Using questionnaires

There are three basic types of questionnaire – **closed-ended**, **open-ended** or a **combination of both**.

1. Closed-ended questionnaires

Closed-ended questionnaires are probably the type with which you are most familiar. Most people have experience of lengthy consumer surveys that ask about your shopping habits and promise entry into a prize draw. This type of questionnaire is used to generate statistics in quantitative research. As these questionnaires follow a set format, and as most can be scanned straight into a computer for ease of analysis, greater numbers can be produced.

2. Open-ended questionnaires

Open-ended questionnaires are used in qualitative research, although some researchers will quantify the answers during the analysis stage (see Chapter 12). The questionnaire does not contain boxes to tick, but instead leaves a blank section for the respondent to write in an answer.

Whereas closed-ended questionnaires might be used to find out how many people use a service, open-ended questionnaires might be used to find out what people think about a service. As there are no standard answers to these questions, data analysis is more complex. Also, as it is opinions that are sought rather than numbers, fewer questionnaires need to be distributed.

3. Combination of both

Many researchers tend to use a combination of both open and closed questions. That way, it is possible to find out how many people use a service and what they think about that service on the same form. Many questionnaires begin with a series of closed questions, with boxes to tick or scales to rank, and then finish with a section of open questions for more detailed response.

Increasingly, market research and opinion poll companies distribute their questionnaires online and pay respondents for their answers. This enables them to build up a following of loyal respondents to whom they can send questionnaires quickly and simply, and receive responses back within shorter deadlines and without the need to pay for postage or send reminder letters. However, in this type of online research the participants are self-selecting, that is they have chosen to take part in the research on a voluntary basis. Some have done this because they are going to be paid and need the money, some enjoy completing questionnaires and others may have a particular axe to grind. If you choose to use this type of 'self-selecting sample' you must be aware of the biases that can occur and be cautious when making generalisations about your findings. More information about these issues is provided in Chapter 5.

TIP

Visit *www.yougov.co.uk* to see an example of a professional research company that uses the internet to collect in-depth data for market and organisational research.

Undertaking participant observation

There are two main ways in which researchers observe: direct observation and participant observation. Direct observation tends to be used in areas such as health and psychology. It involves the observation of a 'subject' in a certain situation and often uses technology such as visual recording equipment or one-way mirrors. For example, the interaction of mother, father and child in a specially prepared play room may be watched by psychologists through a one-way mirror in an attempt to understand more about family relationships.

In participant observation, however, the researcher becomes much more involved in the lives of the people being observed. Using the same research topic described above, a participant observer would not observe the family from a distance. Instead, she would immerse herself in the life of the family in an attempt to understand more about family relationships.

Participant observation can be viewed as both a method and a methodology (see Chapter 10). It is popular among anthropologists and sociologists who wish to study and understand another community, culture or context. They do this by immersing themselves within that culture. This may take months or years, as they need to build up a lasting and trusting relationship with those people being studied. Through participation within their chosen culture and through careful observation, they hope to gain a deeper understanding of the behaviour, motivation and attitudes of the people under study.

Participant observation, as a research method, received bad press when a number of researchers became **covert** participant observers; entering organisations and participating in their activities without anyone knowing that they were conducting research (see Chapter 14). **Overt** participant observation, where everyone knows who the researcher is and what she is doing, however, can be a valuable and rewarding method for qualitative inquiry.

Conducting experiments

There are three main types of experiment involving human beings that are conducted by social researchers: controlled experiments, field experiments and natural experiments.

Controlled experiments

Controlled experiments are carried out in a laboratory and are a systematic, highly focused way to collect data. They tend to be used in medicine, biology, chemistry and engineering, but can also be used in psychology and sociology.

Controlled experiments need an experimental group and a control group. The experimental group is exposed to the factor that is being tested, whereas the control group is not. The factor that is being tested is called the independent variable (or manipulated variable) and the factor that is measured and/or observed is the dependent variable. All other influences on the groups are carefully controlled and are referred to as controlled or constant variables.

Field experiments

Field experiments take place in real environments, such as schools, workplaces and youth clubs, rather than in a laboratory. They tend to be used in political science, economics, sociology, education and psychology. The researcher still controls the independent variable, but tends to have less control over other factors (or variables) that may influence the research.

Some people believe that field experiments better reflect everyday life because they are carried out in everyday settings, rather than in a laboratory. However, because the researcher has less control on external influences, this type of study is harder to replicate (this is important as it helps to show that the research has been conducted properly and is reliable: see Chapter 11).

Natural experiments

Natural experiments, as with field experiments, are carried out in everyday settings. However, they differ from field experiments in that the researcher has no control over the independent variable. They tend to be used in psychology, education, sociology, health care and medicine.

Natural experiments are used when it is not possible, or it is ethically unacceptable, to choose and manipulate the independent variable – for example, in cases of disease outbreak or with research on stress or violence in the home. These experiments can be time-consuming and expensive to conduct, but provide the opportunity to investigate everyday situations that cannot be created in the lab. Again, natural experiments may be difficult to replicate.

> **TIP**
>
> If you intend to conduct experiments, think about how your presence, behaviour and expectations could influence the reactions or behaviour of participants. This can be referred to as 'experimenter-expectancy effect' or 'observer effect', for example.

Choosing your methods

By now you should have thought quite seriously about your research methodology. This will help you to decide upon the most appropriate methods for your research. For example, if you're leaning towards quantitative research, survey work in the form of a questionnaire or structured interviews may be appropriate. If you're interested in action research, it might be useful to find out more about semi-structured interviewing or focus groups.

In quantitative research you can define your research methods early in the planning stage. You know what you want to find out and you can decide upon the best way to obtain the information. Also, you will be able to decide early on how many people you need to contact (see Chapter 5).

However, in some types of qualitative research it may be difficult to define your methods specifically. You may decide that semi-structured interviews would be useful, although you're not sure, in the planning stages, how many you will need to conduct. You may find also that you need to use other methods as the research progresses. Maybe you want to run a focus group to see what people think about the hypotheses you have generated from the interviews. Or perhaps you need to spend some time in the field observing something that has arisen during the interview stage.

Defining needs and means

It is not necessary to use only one research method, although many projects do this. A combination of methods can be desirable, as it enables you to overcome the different weaknesses inherent in all methods. What you must be aware of, however, when deciding upon your methods, are the constraints under which you will have to work. What is your time scale? What is your budget? Are you the only researcher, or will you have others to help you? There's no point deciding that a large-scale, national postal survey is the best way to do your research if you only have a budget of £50 and two months in which to complete your work.

Thinking about purpose

Also, you need to think about the purpose of your research as this will help point to the most appropriate methods to use. For example, if you want to describe in detail the experiences of a group of women trying to set up and run a charity, you wouldn't send them a closed-ended questionnaire. Instead, you might ask to become involved and set up a piece of action research in which you can decide to use interviews and focus groups. Or you might decide to hold two semi-structured interviews with each of the women involved, one at the beginning of their project and one at the end. If your goal is detailed description, you do not need to try to contact as many people as possible.

Let us return to the three examples in the exercises given in the previous two chapters to find out which would be the most appropriate methods for the research.

EXAMPLE 3: APPROPRIATE METHODS

Revised statement 1: *This research aims to find out what primary school teachers think about the educational value of the* Teletubbies *television programme.*

This researcher is interested in attitude and opinion. She thinks about running a series of semi-structured interviews with a small sample of primary school teachers. However, the researcher is concerned that some of the teachers may not have seen the programme and might be unable to comment, or might comment purely on 'hearsay'. So she decides to gather together a group of teachers and show them one episode of *Teletubbies*. Then she discusses the programme with the teachers in a focus group setting. This method works well and the researcher decides to hold five more focus groups with other primary school teachers.

Revised statement 2: *The aim of this research is to find out how many relatives of Alzheimer's patients use the Maple Day Centre, and to ascertain whether the service is meeting their needs.*
This researcher decides to produce a questionnaire with a combination of closed- and open-ended questions. The first part of the questionnaire is designed to generate statistics and the second part asks people for a more in-depth opinion. He has approached members of staff at the Maple Day Centre, who are happy to distribute his questionnaire over a period of one month.

Revised statement 3: *This research aims to find out how many people from our estate are interested in, and would use, a children's play scheme in the school summer holiday.*
Members of the tenants' association approach the local school and ask the head teacher if a questionnaire could be distributed through the school. The head teacher feels that it is not appropriate so the tenants' association have to revise their plans. They're worried that if they distribute a questionnaire through the post they won't receive back many responses. Eventually, they decide to knock on each door on the estate and ask some simple, standard questions. They're able to conduct this type of door-to-door, structured interview as they are a large group and are able to divide the work among everybody on the committee.

If, at this stage, you are still unsure of the most appropriate methods for your research, read the following chapters, as these explain in more detail how to go about using each method. This will give you more of an insight into what would be required of you if you were to choose that method.

As I stressed earlier, you need to think about your own personality, your strengths and weaknesses, your likes and dislikes. If you're a nervous person who finds it difficult to talk to strangers, face-to-face interviewing might not be the best method for you. If you love working with groups, you might like to find out more about focus group research. If a particular culture has fascinated you for years and you know you could immerse yourself within that culture, perhaps participant observation would interest you. If you love number crunching or using statistical software, a closed-ended questionnaire may be the best method for you. Or, if you like to follow set rules and procedures, adhering to the scientific method, then conducting experiments may be the way forward.

TIP

Remember to think about choosing a method or method(s) with which you are happy, as this is important to keep your motivation levels high.

Summary

- Research methods are the tools that are used to gather data.

- Three types of interview are used in social research:
  - Unstructured or life history interviews.
  - Semi-structured interviews.
  - Structured interviews.

- Interviews can be conducted face-to-face or over the telephone.

- Focus groups are held with a number of people to obtain a group opinion.

- Focus groups are run by a moderator who asks questions and makes sure the discussion does not digress.

- Questionnaires can be closed-ended, open-ended or a combination of both.

- Participant observation is used when a researcher wants to immerse herself in a specific culture to gain a deeper understanding.

- Experiments undertaken by social researchers include controlled experiments in a laboratory, field experiments and natural experiments. All are structured and highly focused.

- The chosen research methodology should help to indicate the most appropriate research tools.
- Research methods must be chosen within budget and time constraints.
- The purpose of the research will provide an indicator to the most appropriate methods.
- You should think about your personality, strengths and weakness, likes and dislikes, when choosing research methods.

Further reading

Balnaves, M. and Caputi, P. (2001) *Introduction to Quantitative Research Methods: An Investigative Approach*, London: Sage.

Bickman, L. and Rog, D. (eds) (2008) *The Sage Handbook of Applied Social Research Methods*, 2nd edition, Thousand Oaks, CA: Sage.

Bryman, A. (2016) *Social Research Methods*, 5th edition, Oxford: Oxford University Press.

Denscombe, M. (2017) *The Good Research Guide: for small-scale social research projects*, 6th edition, Buckingham: Open University Press.

Fowler, F. (2014) *Survey Research Methods*, 5th edition, Thousand Oaks, CA: Sage.

Lune, H. and Berg, B. (2017) *Qualitative Research Methods for the Social Sciences*, 9th edition, Harlow: Pearson.

Mason, J. (2002) *Qualitative Researching*, 2nd edition, London: Sage.

McNeill, P. and Chapman, S. (2005) *Research Methods: Textbook*, London: Routledge.

Robson, C. and McCartan, K. (2016) *Real World Research*, 4th edition, Chichester: John Wiley & Sons.

4
How to Conduct Background Research

Once you have decided upon a research project and you're able to sum up your proposed research in one sentence, it's time to start planning your project. The first thing you need to do is your **background research**. This will help you to become more familiar with your topic and introduce you to any other research that will be of benefit to you when you begin your own project.

Conducting primary and secondary research

There are two types of background research – **primary research** and **secondary research** (see Table 2). Primary research involves the study of a subject through firsthand observation and investigation. This is what you will be doing with your main project, but you may also need to conduct primary research for your background work, especially if you're unable to find any previously published material about your topic. Primary research may come from your own observations or experience, or from the information you gather personally from other people, as the following example illustrates.

> **TIP**
> Note that secondary sources interpret, analyse and critique primary sources. They tell a story one or more steps removed from the original person, time, place or event.

EXAMPLE 4: JENNY

I was interested in looking at truancy in schools. The idea came about from my own personal experience as a teacher. I had noticed how some children didn't fit the classic description of a truant and I wanted to find out more as I thought it might help me to deal with some of the problems children were experiencing. So I guess you'd say my own experience provided me with some initial data.

Then I decided to go and have a discussion with some of my colleagues and see if they'd noticed anything like me. It was really useful to do this because they helped me to think about other things I hadn't even thought of.

One of them told me about a new report which had just come out, and it was useful for me to go and have a look at it as it raised some of the issues I was already thinking about. Actually this made me change the focus of my work a little because I soon found out that there had been a lot of work on one area of what I was doing, but not so much on another area. It was really useful to have done this before I rushed into my research as I think I might otherwise have wasted quite a bit of time.

In the above example, Jenny mentions a recently published report that she has read. This is **secondary research** and it involves the collection of information from studies that other researchers have made of a subject. The two easiest and most accessible places to find this information are libraries and the internet. However, you must remember that anybody can publish information online and you should be aware that some of this information can be misleading or incorrect.

Of course this is the case for any published information, and as you develop your research skills so you should also develop your critical thinking and reasoning skills. Do not believe everything you're told. Think about the information you are being given. How was it collected? Were the methods sound? What motives did the publishers have for making sure their information had reached the public domain?

By developing these skills early in your work, you will start to think about your own research and any personal bias in your methods and reporting that may be present.

Using websites

University websites carry information about how to use the web carefully and sensibly for your research, and it is worth accessing these before you begin your background work. The following points will help you to surf the net effectively and efficiently.

- Try to use websites run by organisations you know and trust.

- Check the *About Us* section on the web page for more information about the creator and organisation.
- Use another source, if possible, to check any information of which you are unsure. For example, if you're interested in medical information you can check the credentials of UK doctors by phoning the General Medical Council.
- You should check the national source of the data, as information may differ between countries.
- For some topics, specific websites have been set up that contain details of questionable products, services and theories. For example, in health and medical research you can visit *www.quackwatch.org*, which provides information about health-related frauds, questionable theories and money-making scams.
- If you come across information that is useful for your research, remember to keep a record of all the information you will need to reference the website in your report or dissertation (see Chapter 13).
- Although it is tempting to cut and paste information from relevant websites, you must be careful if you choose to adopt this procedure. All cut and paste sections must be clearly marked so that you do not accidentally (or intentionally) pass off the information as your own work. Plagiarism is a serious offence and universities may use plagiarism – detecting software to catch students who cheat with their work. However, it may be possible to cut and paste relevant quotations, as long as you use quotation marks in your report and reference the work carefully and correctly. If you choose to do this, you must make sure that you don't breach copyright laws. If in doubt, always seek permission from the author of the website before using a quotation. More information about copyright can be obtained from *www.gov.uk/topic/intellectual-property/copyright*.

TIP

Before you start searching online, define your research topic, research methods and research needs. Think carefully about what sources will provide you with the information you need without overloading you with data.

Using interlibrary loans

If you are a student, your institutional library will probably offer an interlibrary loan service, which means that you can access books from other university libraries if they are not available in your library.

A useful website is *www.copac.jisc.ac.uk*, which provides free access to the online catalogues of university research libraries in the UK and Ireland. This is a useful service if, when referencing, you find that a small amount of information is missing (see Example 5 below).

EXAMPLE 5: GILLIAN

Nobody told me the importance of keeping careful records of my background research. I just thought it was something you did and then that was it, you got on with your own research and forgot about what you'd done. Of course then I had to write my report and in the 'background' section I wanted to include loads of things I'd read when I first started the work. I found my notes, but I didn't know where they'd come from. It was so frustrating. Basically I had to start all over again. Even then I still forgot to write down the name and location of the publisher, so I had to go back to them again. My advice would be to look at how bibliographies are structured and imprint that in your brain so you don't forget anything.

Keeping records

When you begin your background research, keep accurate records of what data was gathered from which source, as this will save you plenty of time and frustration later, especially when you come to write your research proposal, or final report. A useful way to organise your notes is to separate your files and folders into primary and secondary research. For handwritten notes, such as those taken from library books and journal articles, keep an A4 file with the relevant pages slotted into the separate sections. For information stored on your PC or laptop, such as typed notes, transcripts and information obtained from websites, place the documents in the relevant folder and keep document and folder names simple so that you can access the information easily when required. Remember to back up all folders and documents on a regular basis.

Primary research

For the primary research file or folder, notes from each contact can be separated by a contact sheet or document that gives the name of the person, the date and time you met, and a contact number or address.

Secondary research

In the secondary research file or folder, each page of notes or document can be headed by details of the publication in the same format that will be used in the bibliography: author and initials; date of publication; title of publication; place of publication; and publisher. If it is a journal article, remember to include the name of the journal; the page numbers of the article, and the volume and number of the journal. It is also useful to include the location of this publication so that it can be found easily if needed again (website or library shelf location).

There are tools and software available that make citing, referencing and producing a bibliography a simple and quick process. Some of these are freely available online, whereas others will be available from your university or place of work. Popular tools include Cite This For Me (*www.citethisforme.com*), BibMe (*www.bibme.org*), RefWorks (*www.refworks.com*) and EndNote (*endnote.com*). There are plenty more available, including useful apps for mobile devices, so hunt around for the tool that best suits your needs.

TABLE 2: SOURCES OF BACKGROUND INFORMATION

PRIMARY	SECONDARY
Relevant people	Research books
Researcher observation	Research critiques or reviews
Researcher experience	Journal reviews
Historical records/texts	Critiques of literary works
Company/organisation records	Critiques of art
Personal documents (diaries, etc.)	Analyses of historical events
Statistical data	Biographies
Works of literature	Television documentaries
Works of art	Analysis of clinical trials
Film/video	
Laboratory experiments	

Summary

- There are two types of background research: primary and secondary research.
- Primary research involves the study of a subject through firsthand observation and investigation.
- Secondary research involves the collection of information from studies that other researchers have made of a subject.
- For most research, the easiest and quickest way to access secondary sources are libraries or the internet.
- Any information obtained from secondary sources must be assessed carefully for its relevance and accuracy.
- Notes from primary and secondary sources should be carefully filed and labelled so that the source can be found again, if required.
- When noting details for books, reports or articles that may appear in the final report, include all the details that would be needed for the bibliography.

Further reading

Boland, A., Cherry, G. and Dickson, R. (2013) *Doing a Systematic Review*, London: Sage.

Booth, A., Sutton, A. and Papaioannou, C. (2016) *Systematic Approaches to a Successful Literature Review*, London: Sage.

Hart, C. (2001) *Doing a Literature Search*, London: Sage.

Machi, L. and McEvoy, B. (2016) *The Literature Review: Six Steps to Success*, Thousand Oaks, CA: Corwin.

5
How to Choose Your Participants

As you continue planning your research project, you need to think about how you're going to choose your participants. By now you should have decided what type of people you need to contact. For some research projects, there will be only a small number of people within your research population, in which case it might be possible to contact everyone. This is called a **census**. However, for most projects, unless you have a huge budget, limitless time scale and a large team of interviewers, it will be difficult to speak to every person within your research population.

Understanding sampling techniques

Researchers overcome this problem by choosing a smaller, more manageable number of people to take part in their research. This is called **sampling**. In quantitative research, it is believed that if this sample is chosen carefully using the correct procedure, it is then possible to generalise the results to the whole of the research population.

For many qualitative researchers, however, the ability to generalise their work to the whole research population is not the goal. Instead, they might seek to describe or explain what is happening within a smaller group of people. This, they believe, might provide insights into the behaviour of the wider research population, but they accept that everyone is different and that if the research were to be conducted with another group of people the results might not be the same.

Using sampling procedures

Sampling procedures are used every day. Market researchers use them to find out what the general population think about a new product or new advertisement. When they report that 87 per cent of the population like the smell of a new brand of washing powder, they haven't spoken to the whole population, but instead have contacted only a sample of people who they believe are able to represent the whole population.

When we hear that 42 per cent of the population intend to vote Labour at the next General Election, only a sample of people have been asked about their voting intentions. If the sample has not been chosen very carefully, the results of such surveys can be misleading. Imagine how misleading the results of a 'national' survey on voting habits would be if the interviews were conducted only in the leafy suburbs of an English southern city.

Probability samples and purposive samples

There are many different ways to choose a sample, and the method used will depend upon the area of research, research methodology and preference of the researcher. Basically there are two main types of sample:

- probability samples
- purposive samples.

In probability samples, all people within the research population have a specifiable chance of being selected. These types of sample are used if the researcher wishes to explain, predict or generalise to the whole research population. On the other hand, purposive samples are used if description rather than generalisation is the goal. In this type of sample it is not possible to specify the possibility of one person being included in the sample. Within the probability and purposive categories there are several different sampling methods.

The best way to illustrate these sampling methods is to take one issue and show how the focus of the research and the methodology leads to the use of different sampling methods. The area of research is 'school detention' and in Table 3 you can see that the focus and sampling techniques within this topic can be very different, depending on the preferences of the researcher, the purpose of the research and the available resources.

TABLE 3: SAMPLING TECHNIQUES

PROBABILITY SAMPLES	**PURPOSIVE SAMPLES**
The researcher is interested in finding out about national detention rates. He wants to make sure that every school in the country has an equal chance of being chosen, because he hopes to be able to make generalisations from his findings. He decides to use a **simple random sample**. Using this method the researcher needs to obtain the name of every school in the country. Numbers are assigned to each name and a random sample generated by computer. He then sends a questionnaire to each of the selected schools. The researcher would have to make sure that he obtained the name of every school in the country for this method to work properly.	The researcher decides that he wants to interview a sample of all pupils within a school, regardless of whether they have been on detention or not. He decides to use a **quota sample** to make sure that all groups within the school are represented. He decides to interview a specified number of female and male school pupils, a specified number of arts, sciences and social science pupils, and a specified number within different age categories. He continues approaching students and interviewing them until his quota is complete. By using this method, only those pupils present at the same time and in the same place as the researcher have a chance of being selected.
The researcher wants to find out about national detention rates, but is interested also in finding out about school policy concerning detention. He decides that to do this he needs to visit each selected school. To cut down on travel costs, he decides to use a **cluster sample**. Using this method, geographical 'clusters' are chosen and a random sample of schools from each cluster is generated using random number tables found at the back of some statistics books. Using this method the researcher only needs to travel to schools within the selected geographical regions. The researcher would have to make sure that he chose his clusters very carefully, especially as policy concerning detention might vary between regions.	The researcher is interested in carrying out semi-structured interviews with pupils who have been on detention over the past year. However, he finds that the school has not kept accurate records of these pupils. Also, he doesn't want to approach the school because he will be seen by the pupils as an authority figure attached to the school. He decides that a **snowball sample** would be the most appropriate method. He happens to know a pupil who has been on detention recently and so speaks to her, asking for names of other pupils who might be willing to talk to him. The researcher should obtain permission and have a chaperone or guardian present at the interviews. He needs to be aware also that friends tend to recommend friends, which could lead to sampling bias.

PROBABILITY SAMPLES	PURPOSIVE SAMPLES
The researcher has decided that he wishes to conduct a structured interview with all the children who have been on detention within a year at one school. With the head teacher's permission, he obtains a list of all these pupils. He decides to use a **quasi-random sample** or **systematic sample**. Using this method he chooses a random point on the list and then every third pupil is selected. The problem with this method is that it depends upon how the list has been organised. If, for example, the list has been organised alphabetically, the researcher needs to be aware that some cultures and nationalities may have family names that start with the same letters. This means that these children would be grouped together in the list and may, therefore, be under-represented in the sample.	The researcher has heard of a local school which has very few detentions, despite that school having a detention policy. He decides to find out why and visits the school to speak to the head teacher. Many interesting points arise from the interview and the researcher decides to use a **theoretical sampling** technique. Using this method the emerging theory helps the researcher to choose the sample. For example, he might decide to visit a school that has a high detention rate and a school that has no detention policy, all of which will help to explain differing detention rates and attitudes towards them. Within this sampling procedure, he might choose to sample **extreme cases** that help to explain something, or he might choose **homogeneous samples** where there is a deliberate strategy to select people who are alike in some relevant detail. Again the researcher has to be aware of sampling bias.
The researcher has decided that he wishes to concentrate on the detention rates of pupils by GCSE subject choice and so decides upon a **stratified random sample**. Using this method the researcher stratifies his sample by subject area and then chooses a random sample of pupils from each subject area. However, if he found that there were many more pupils in the arts than the sciences, he could decide to choose a **disproportionate stratified sample** and increase the sample size of the science pupils to make sure that his data are meaningful. The researcher would have to plan this sample very carefully and would need accurate records of subjects and pupils.	The researcher is a teacher himself and decides to interview colleagues, as he has limited time and resources available to him. This is a **convenience sample**. Also, at a conference, he unexpectedly gets to interview other teachers. This might be termed **haphazard or accidental sampling**. The ability to generalise from this type of sample is not the goal, and, as with other sampling procedures, the researcher has to be aware of bias that could enter the process. However, the insider status of the teacher may help him to obtain information or access that might not be available to other researchers.

TABLE 4: SAMPLING DOS AND DON'TS

DO	DON'T
Take time and effort to work out your sample correctly if you're conducting a large-scale survey. Read the relevant literature suggested in this book. Time taken at the beginning will save much wasted time later.	Rush into your work without thinking very carefully about sampling issues. If you get it wrong it could invalidate your whole research.
Discuss your proposed sampling procedure and size with your tutor, boss or other researchers.	Ignore advice from those who know what they're talking about.
Be realistic about the size of sample possible on your budget and within your time scale.	Take on more than you can cope with. A badly worked-out, large sample may not produce as much useful data as a well-worked-out, small sample.
Be open and up front about your sample. What are your concerns? Could anything have been done differently? How might you improve upon your methods?	Make claims that cannot be justified or generalised to the whole population.
Use a combination of sampling procedures if it is appropriate for your work.	Stick rigorously to a sampling technique that is not working. Admit your mistakes, learn by them and change to something more appropriate.

Remaining cautious

Even though you may be restricted by time and finance, you should be wary of convenience samples in quantitative research. For example, it is becoming increasingly popular for researchers to use web-based surveys, but in many of these the volunteers are self-selecting. People have many different reasons for putting themselves forward – perhaps they have an axe to grind, perhaps they are 'serial respondents'. Whatever the reason, these people may not be representative of the views and attitudes of the groups of people to whom you hope to apply your research. In these cases you will not be able to make generalisations.

Choosing your sample size

The first question new researchers tend to ask is 'how many people should I speak to?' This obviously depends on the type of research. For large-scale quantitative surveys you will need to contact many more people than you would for a small, qualitative piece of research. The sample size will also depend on what you want to do with your results. If you intend to produce large amounts of cross tabulations, the more people you contact the better.

It tends to be a general rule in quantitative research that the larger the sample the more accurate your results. However, you have to remember that you are probably restricted by time and money – you have to make sure that you construct a sample that will be manageable. Also, you have to account for non-response and you may need to choose a higher proportion of your research population as your sample to overcome this problem. If you're interested in large-scale quantitative research, statistical methods can be used to choose the size of sample required for a given level of accuracy and the ability to make generalisations. These methods and procedures are described in the statistics books listed at the end of this chapter.

If your research requires the use of purposive sampling techniques, it may be difficult to specify at the beginning of your research how many people you intend to contact. Instead you continue using your chosen procedure, such as snowballing or theoretical sampling, until a 'saturation point' is reached. This was a term used by Glaser and Strauss (1967) to describe that time of your research when you really do think that everything is complete and that you're not obtaining any new information by continuing. In your written report you can then describe your sampling procedure, including a description of how many people were contacted.

TIP

Note that 'sample size' refers to the number of individuals or groups that are required to respond to achieve the required level of accuracy. It concerns the final number of responses collected, rather than the number of individuals selected to provide responses. Potential response rates will need to be considered.

Summary

- If it is not possible to contact everyone in the research population, researchers select a number of people to contact. This is called sampling.
- There are two main types of sampling category – probability samples and purposive samples.
- In probability samples, all people within the research population have a specifiable chance of being selected. Only within random samples do participants have an equal chance of being selected.
- Purposive samples are used if generalisation is not the goal.
- The size of sample will depend upon the type and purpose of the research.
- Sample sizes should take into account issues of non-response.
- Remember that with postal surveys it might be difficult to control and know who has filled in a questionnaire. Will this affect your sample?
- In some purposive samples it is difficult to specify at the beginning of the research how many people will be contacted.
- It is possible to use a mixture of sampling techniques within one project, which may help to overcome some of the disadvantages found within different procedures.

Further reading

De Vaus, D. (2014) *Surveys in Social Research*, 6th edition, London: Routledge.

Emmel, N. (2013) *Sampling and Choosing Cases in Qualitative Research*, London: Sage.

Fink, A. (2016) *How to Conduct Surveys: A Step by Step Guide*, 6th edition, Thousand Oaks, CA: Sage.

Fowler, F.J. (2014) *Survey Research Methods*, 5th edition, Thousand Oaks, CA: Sage.

Henry, G. (1990) *Practical Sampling*, Newbury Park, CA: Sage.

Huff, D. (1994) *How to Lie With Statistics*, New York: Norton.

6
How to Prepare a Research Proposal

For most types of research you will need to produce a **research proposal**. This is a document that sets out your ideas in an easily accessible way. Even if you have not been asked specifically to produce a research proposal by your boss or tutor, it is a good idea to do so, as it helps you to focus your ideas and provides a useful document for you to reference, should your research wander off track a little.

Understanding the format

Before you start work on your research proposal, find out whether you're required to produce the document in a specific format. For college and university students, you might be given a general outline and a guide as to how many pages to produce. Make sure you familiarise yourself with structures, rules and regulations before you begin your work.

For those of you who are producing a proposal to send to a funding organisation, you might have to produce something much more specific. Many funding organisations provide their own forms for you to complete. Some provide advice and guidance about what they would like to see in your proposal. The larger funding bodies provide their proposal forms online so that they can be filled in and sent electronically, which makes the process a lot quicker and easier.

The contents of a proposal

All research proposals should contain the following information:

Title

This should be short and explanatory.

Background

This section should contain a rationale for your research that answers the following questions:

- Why are you undertaking the project?
- Why is the research needed?

This rationale should be placed within the context of existing research or within your own experience and/or observation. You need to demonstrate that you know what you're talking about and that you have knowledge of the literature surrounding this topic.

If you're unable to find any other research that deals specifically with your proposed project, you need to say so, illustrating how your proposed research will fill this gap. If there is other work that has covered this area, you need to show how your work will build on and add to the existing knowledge. Basically, you have to convince people that you know what you're talking about and that the research is important.

Aims and objectives

Many research proposal formats will ask for only one or two aims and may not require objectives. However, for some research these will need to be broken down in more depth to also include the objectives (see Example 6).

The aim is the overall driving force of the research and the objectives are the means by which you intend to achieve the aims. These must be clear and succinct.

EXAMPLE 6: AIMS AND OBJECTIVES

Aim

To identify, describe and produce an analysis of the interacting factors which influence the learning choices of adult returners, and to develop associated theory.

Objectives

The research seeks to determine:

1. The nature, extent and effect of psychological influences on choices, including a desire to achieve personal goals or meet individual needs.

2. The nature, extent and effect of sociological influences on choices, including background, personal and social expectations, previous educational experience and social role.

3. The nature and influence of individual perceptions of courses, institutions and subject, and how these relate to self-perception and concept of self.

4. The influence on choice of a number of variables such as age, gender, ethnicity and social class.

5. The role and possible influence of significant others on choice, such as advice and guidance workers, peers, relatives and employers.

6. The nature and extent of possible influences on the choice of available provision, institutional advertising and marketing.

7. The nature and extent of possible influences on the choice of mode of study, teaching methods and type of course.

8. How and to what extent influencing factors change as adults re-enter and progress through their chosen route.

Methodology/methods

For research at postgraduate level you may need to split the methodology and methods section into two. However, for most projects they can be combined. In this section you need to describe your proposed research methodology and methods and justify their use. To do this you need to ask the following questions:

- Why have you decided upon your methodology?
- Why have you decided to use those particular methods?
- Why are other methods not appropriate?

This section needs to include details about samples, numbers of people to be contacted, method of data collection, methods of data analysis, and ethical considerations. If you have chosen a less well-known methodology, you may need to spend more time justifying your choice than you would need to if you had chosen a more traditional methodology.

This section should be quite detailed – many funding organisations find that the most common reason for proposal failure is the lack of methodological detail (see below).

Timetable

A detailed timetable scheduling all aspects of the research should be produced. This will include time taken to conduct background research, questionnaire or interview schedule development, data collection, data analysis and report writing (see Table 5).

Research almost always takes longer than you anticipate. Allow for this and add a few extra weeks on to each section of your timetable. If you finish earlier than you anticipated, that's fine, as you have more time to spend on your report. However, finishing late can create problems, especially if you have to meet deadlines.

Budget and resources

If you're applying to a funding body, you need to think about what you will need for your research and how much this is likely to cost (see Table 6). You need to do this so that you apply for the right amount of money and are not left out of pocket if you have under-budgeted. Funding bodies also need to know that you have not over-budgeted and expect more money than you're going to use.

If you are a student, you may not have to include this section in your proposal, although some tutors will want to know that you have thought carefully about what resources are needed and from where you expect to obtain these.

Some types of research are more expensive than others and if you're on a limited budget you will have to think about this when deciding upon your research method.

TABLE 5: SURVEY TIMETABLE

DATE	ACTION
5 January – 5 February	Literature search Primary research (talk to relevant people)
6 February – 7 March	Develop and pilot questionnaire Continue literature search
8 March – 9 April	Analyse pilot work and revise questionnaire Ask relevant people for comments
10 April – 21 April	Send out questionnaire Categorise returned questionnaires
21 April – 1 May	Send out reminder letter for non-responses. Continue to categorise returned questionnaires
1 May – 1 July	Data input Data analysis
2 July – 3 August	Write report Prepare oral presentation

TABLE 6: RESEARCH BUDGET

RESOURCE	COST
1 good-quality digital voice recorder	£31.99
20 long-life batteries	£9.99
40 second-class postage stamps	£22.40
Stationery – paper, envelopes, paper clips, ring binder, scissors	£8.76
Travel expenses – petrol, overnight stay at five locations	Petrol to be notified at usual college mileage allowance Total accommodation £199.95
Advert in local paper	£3.70
Leaflets (1,000)	£21.90
Total Expenditure	£298.69 + petrol (to be notified)

Dissemination

What do you expect to do with the results of your research? How are you going to let people know about what you have found out? For students it will suffice to say that the results will be produced in an undergraduate dissertation that will be made available in the institution library. For other researchers you may want to produce a written report, make oral presentations to relevant bodies, produce a website, write a journal article or produce a series of recommendations to aid decision making.

What makes a good proposal?

- Relevance, either to the work of the funding body or to the student's course.
- The research is unique, or offers new insight or development.
- The title, aims and objectives are all clear and succinct.
- Comprehensive and thorough background research and literature review has been undertaken.
- There is a good match between the issues to be addressed and the approach being adopted.
- The researcher demonstrates relevant background knowledge and/or experience.
- Timetable, resources and budget have all been worked out thoroughly, with most eventualities covered.
- Useful policy and practice implications.

Reasons why research proposals fail

- Aims and objectives are unclear or vague.
- There is a mismatch between the approach being adopted and the issues to be addressed.
- The overall plan is too ambitious and difficult to achieve in the time scale.
- The researcher does not seem to have conducted enough in-depth background research.
- Problem is of insufficient importance.
- Information about the data collection method is insufficiently detailed.
- Information about the data analysis method is insufficiently detailed.
- Time scale is inappropriate or unrealistic.
- Resources and budget have not been thought out carefully.
- This topic has been done too many times before – indicates a lack of background research.

TIP

Consult your university website or ask your tutor for examples of successful research proposals. If you are applying to a funding organisation, ask for examples of proposals that were successful in receiving their funding. Analyse the proposals, taking note of the good, effective points. Once you have produced your own proposal, ask friends or colleagues to read it through and suggest improvements that could be made.

Summary

- Most research projects will require the production of a research proposal that sets out clearly and succinctly your proposed project.

- Before you write your proposal, check whether you need to produce it in a specific format.

- The standard research proposal should include the following:
  - title
  - background (including literature search)
  - aims and objectives
  - methodology/methods
  - timetable
  - budget and resources
  - dissemination.

- Research proposals stand a better chance of being accepted if you're able to prove that you have the required knowledge and/or experience to carry out the research effectively.

- It is important to make sure that your proposed methods will address the problem you have identified and that you are able to display an understanding of these methods.

Further reading

Dawson, C. (2015) *How to Finance your Research Project: A Practical Guide to Costing Projects and Obtaining Funding*, London: Robinson.

Locke, L.F., Spirduso, W.W. and Silverman, S.J. (2013) *Proposals That Work: A Guide for Planning Dissertations and Grant Proposals*, 6th edition, Thousand Oaks, CA: Sage.

Punch, K.F. (2006) *Developing Effective Research Proposals*, 2nd edition, London: Sage.

7
How to Conduct Interviews

As we have seen in Chapter 3, there are three main types of interview that tend to be used in social research – unstructured interviews, semi-structured interviews and structured interviews. For each type you will need to think about how you are going to record the interview, what type of questions you need to ask, how you intend to establish rapport and how you can probe for more information.

Methods of recording

If you've decided that interviewing is the most appropriate method for your research, you need to think about what sort of recording equipment you're going to use. The advantages and disadvantages of each method are listed in Table 7. You should think about recording methods early on in your research, as you need to become familiar with their use through practice. Even if you decide not to use audio recording equipment, and instead use pen and paper, you should practise taking notes in an interview situation, making sure that you can maintain eye contact and write at the same time.

If, however, you're conducting a structured interview, you will probably develop a questionnaire with boxes to tick as your method of recording (see Chapter 9). This is perhaps the simplest form of recording, although you will have to be familiar with your questionnaire, to make sure you can work through it quickly and efficiently.

Using audio recording equipment

If you are a student, find out what audio recording equipment is available for your use. Today there is a wide variety of recorders available – prices vary enormously and some designs and types are more suited to certain tasks than others. The general rule is that the more you pay for your equipment the better the recording will be. However, there are many other factors that influence the

TABLE 7: RECORDING METHODS: ADVANTAGES AND DISADVANTAGES

RECORDING METHOD	ADVANTAGES	DISADVANTAGES	ADDITIONAL INFORMATION
Audio recording equipment	• Can concentrate on listening to what the interviewee says. • Able to maintain eye contact. • Have a complete record of interview for analysis, including what is said and interaction between interviewer and interviewee. • Have plenty of useful quotations for report.	• Relies on equipment – if it fails you have no record of interview. • Can become complacent – don't listen as much as you should because it's being recorded. • Some interviewees may be nervous of recording equipment.	• Overcome equipment failure by practice beforehand and checking throughout interview, without drawing attention to machine. • Could take a few notes as well – this helps you to write down important issues and you will have some record if equipment fails.
Visual recording equipment	• Produces the most comprehensive recording of an interview. • Gives a permanent record of what is said and includes a record of body language, facial expressions and interaction.	• The more equipment you use the more chances there are that something will go wrong. • This method can be expensive and the equipment hard to transport. • Some interviewees may be nervous of visual recording equipment.	• If you want to use visual recording equipment, it is preferable to obtain the help of someone experienced in the use of the equipment. That way you can concentrate on the interview while someone else makes sure that it is recorded correctly.
Note-taking (paper or laptop)	• You don't have to rely on recording equipment, which could fail. • Is the cheapest method if on a very limited budget. • Interviewees may think they have something important to say if they see you taking notes – while you write, they may add more information.	• Cannot maintain eye contact all the time. • Can be hard to concentrate on what the interviewee is saying and to probe for more information. • Can be tiring. • Will not have many verbatim quotations for final report.	• Need to make sure that you have a suitable venue for this type of recording. • You will need to develop a type of shorthand that you can understand and you will need to learn to write or type very quickly.
Box-ticking	• Simple to use. • Easy to analyse. • Easy to compare information with that obtained from other interviews.	• Inflexible – no scope for additional information. • Forces interviewees to answer in a certain way. • May leave interviewees feeling that they have not answered in the way they would have liked to have done.	• You have to make sure that the questionnaire is very carefully designed so that you cover as many types of answer as possible.

quality of your recording, and even if you are on a limited budget you can still obtain a good recording using cheap equipment if you plan carefully.

RECORDING CHECKLIST

- Make sure that there is enough space (memory, disk or tape) to record everything you need.
- Check that the batteries are fully functioning and that you have enough power to complete your recording. A battery indicator light is useful so that you can check discreetly that you are still recording.
- Do you have spare disks, memory or cassettes, if required?
- Is the venue free from background noise that could disrupt the recording?
- Is there a suitable surface on which to place the equipment?
- Is the microphone strong enough to pick up all voices in a focus group setting? Although built-in microphones are more convenient, you will need to test that they are sufficiently powerful, and if not you may need to consider an external microphone. Built-in microphones tend to be suitable for one-to-one interviews, as long as you are both sitting close enough.
- Are you able to download files to your PC? Do you have the necessary software available? If you can't download files, how easy will it be to transcribe your interview?
- Make sure that storage is safe and secure and that data, disks or cassettes are not lost or accidentally wiped.

It is useful to take a pen and notepad with you to the interview, even if you intend to use a recorder. You might find it useful to jot down pertinent points to which you want to return later, or use it to remind yourself of what you haven't yet asked. Also, you might encounter someone who doesn't want to be recorded. This could be because the research is on a sensitive issue, or it might be that the interviewee has a fear of being recorded.

Recording sensitive topics

In cases where your interview touches on sensitive topics you might find it advantageous to offer to switch off your recording equipment. Often this will spur on a hesitant interviewee to make comments that they might not have done otherwise. You will also need to assure them that all information they supply, whether recorded or not, will remain confidential (see Chapter 14).

Taking notes

If you intend to take paper notes, buy yourself a shorthand notepad and develop a shorthand style that you'll be able to understand later (see Chapter 10). If you intend to use a laptop, make sure that you can type quickly and accurately. It is advisable to write up all notes into a longer report as soon as possible after the interview while it's still fresh in your mind.

If you can type quicker than you can write, and you have use of a laptop, it is more efficient to type your notes as the interviewee speaks, although you should check that they are happy for you to do this, as some people may find it a little off-putting. An advantage to taking notes in this way is that you can organise your notes easily and cut and paste relevant information and quotations into documents and reports.

It can be tiring taking notes in long interviews, so only arrange one or two per day. You must learn to try to maintain some eye contact while you're writing or typing, and make sure that you nod every now and again to indicate that you're still listening. Try also to get one or two verbatim quotations, as these will be useful for your final report.

Developing an interview schedule

For most types of interview you need to construct an **interview schedule**. For structured interviews you will need to construct a list of questions that are asked in the same order and format to each participant (see Chapter 9). For semi-structured interviews the schedule may be in the form of a list of questions or of topics.

If you're new to research, you might prefer a list of questions that you can ask in a standard way, thus ensuring that you do not ask leading questions or struggle for something to ask.

However, a list of topics tends to offer more flexibility, especially in unstructured interviews where the interviewee is left to discuss issues she deems to be important. By ticking off each topic from your list as it is discussed, you can ensure that all topics have been covered. Often interviewees will raise issues without being asked and a list of topics ensures that they do not have to repeat themselves. Also, it allows the interviewee to raise pertinent issues that you may not have thought about. These can then be added to the schedule for the next interview.

Overcoming nerves

If you're nervous about working with a list of topics rather than a list of questions, a good way to overcome this is to ask a few set questions first and then, once you and the interviewee have both relaxed, move on to a set of topics. With practice, you will feel comfortable interviewing and will choose the method that suits you best.

Focusing your mind

If you take time to produce a detailed interview schedule, it helps you to focus your mind on your research topic, enabling you to think about all the areas that need to be covered. It should also alert you to any sensitive or controversial issues that could arise.

When developing an interview schedule for any type of interview, begin with easy-to-answer, general questions that will help the interviewee feel at ease. Don't expect in-depth, personal disclosure immediately.

HOW TO DEVELOP AN INTERVIEW SCHEDULE

- Brainstorm your research topic – write down every area you can think of without analysis or judgement.
- Work through your list carefully, discarding irrelevant topics and grouping similar suggestions.

- Categorise each suggestion under a list of more general topics.

- Order these general topics into a logical sequence, leaving sensitive or controversial issues until the end – ask about experience and behaviour before asking about opinion and feelings. Move from general to specific.

- Think of questions you will want to ask relating to each of these areas. If you're new to research, you might find it useful to include these questions on your schedule. However, you do not have to adhere rigidly to these during your interview.

- When developing questions, make sure they are open rather than closed. Keep them neutral, short and to the point. Use language that will be understood. Avoid jargon and double-barrelled questions (see Chapter 9).

- If you need to, revise your schedule after each interview.

- Become familiar with your schedule so that you do not have to keep referring to it during the interview.

TIP

If you are new to interviewing, produce your interview schedule and practise your interview technique with a trusted friend. Record the interview in the same way that you intend to record subsequent interviews. Ask your friend to make comments about your technique and listen to your recording to see whether there is anything about your technique that could be improved. Practise working through your schedule until you are comfortable and familiar with the topics. If you are comfortable and relaxed, this will rub off on your interviewee.

Establishing rapport

A researcher has to establish **rapport** before a participant will share personal information. There are a number of ways to do this.

- **Treat interviewees with respect.** Make sure you arrive on time. Don't rush straight into the interview unless the interviewee pushes to do so. Accept a cup of tea, if offered, and make polite conversation to help put both of you at ease.

- **Think about your appearance** and the expectations of the person you're about to interview. If the interviewee is a smartly turned-out business person who expects to be interviewed by a professional-looking researcher, make sure you try to fulfil those expectations through your appearance and behaviour.

- **Think about body language.** Try not to come across as nervous or shy. Maintain appropriate eye contact and smile in a natural, unforced manner. Remember that the eyes and smile account for more than 50 per cent of the total communication in a greeting situation. If you establish rapid and clear eye contact, you'll be more easily trusted.

- During the interview, **firm eye contact** with little movement indicates that you're interested in what is being said. Also, it indicates honesty and high self-esteem. On the other hand, if your eyes wander all over the place and only briefly make contact with the eyes of the interviewee, low self-esteem, deceit or boredom can be indicated. Don't rub your eyes, as this could indicate you're tired or bored. Conversely, watch the eyes of your interviewees, which will tell you a lot about how the interview is progressing.

- **Don't invade their space.** Try not to sit directly opposite them – at an angle is better, but not by their side as you will have to keep turning your heads, which will be uncomfortable in a long interview.

By watching the eye movements and body language of the interviewees, and by listening to what they're saying, you'll soon know when you've established rapport. This is when you can move on to more personal or sensitive issues. If, however, you notice the interviewees becoming uncomfortable in any way, respect their feelings and move on to a more general topic. Sometimes you might need to offer to turn off the recorder or stop taking notes if you touch upon a particularly sensitive issue.

Asking questions and probing for information

As the interview progresses, ask questions, listen carefully to responses and probe for more information. You should probe in a way that doesn't influence the interviewee. When you probe, you need to think about obtaining clarification, elaboration, explanation and understanding. There are several ways to probe for more detail, as the following list illustrates. It's useful to learn a few of these before you begin your interviews.

PROBING FOR MORE DETAIL

- That's interesting; can you explain that in more detail?
- I'm not quite sure I understand. You were saying?
- Can you elaborate a little more?
- Could you clarify that?
- Could you expand upon that a little?
- When you say '. . . . . .', what do you mean?

Pauses work well – don't be afraid of silence. You'll find that most people are uncomfortable during silences and will elaborate on what they've said rather than experience discomfort. Also, you may find it helpful to summarise what people have said as a way of finding out if you have understood them and to determine whether they wish to add any further information.

Another useful tactic is to repeat the last few words a person has said, turning it into a question. The following piece of dialogue from an interview illustrates how these techniques can be used so that the researcher does not influence what is being said.

Janet:	'Well, often I find it really difficult because I just don't think the information's available.'
Interviewer:	'The information isn't available?'
Janet:	'No, well I suppose it is available, but I find it really difficult to read, so it makes me think it isn't available.'
Interviewer:	'In what way do you find it difficult to read?'

Janet:	'Well, the language is a bit beyond me, but also the writing's too small and it's a funny colour.'
Interviewer:	'You say the language is a bit beyond you?'
Janet:	'Yes, I suppose really that's why I need to do this, so that it won't be beyond me any more.' [laughs]
Interviewer:	'Why do you laugh about that?'
Janet:	'Well, I don't know, I suppose maybe I'm embarrassed, you know, about not being able to read and write so well, you know, I always blame my eyesight and things being a funny colour and everything, but I suppose the bottom line is I just can't read proper. That's why I'm doing this, you know, going to college and all that. I weren't exactly naughty at school, I just didn't really bother, you know, I didn't really like it that much, if I'm honest with you.'

This piece of dialogue illustrates how, with careful probing, the researcher can discover a greater depth of information that wasn't initially offered by the interviewee.

Completing the interview

Negotiate a length of time for the interviews and stick to it, unless the interviewees are happy to continue. Make sure you thank them for their help and leave a contact number in case they wish to speak to you at a later date.

You might find it useful to send a transcript to the interviewees – it is good for them to have a record of what has been said and they might wish to add further information. Do not disclose information to third parties unless you have received permission to do so (see Chapter 14).

Summary

- Practise with the recording equipment before the interview takes place. It might be useful to conduct some pilot interviews so that you can become familiar with the recording equipment.

- Develop an interview schedule, starting with general, non-personal issues.

- Check the recording equipment works and make sure you have enough tapes, disks, memory and/or batteries, paper, pens, etc.

- Check that you have a suitable venue in which to carry out the interview, free from noise and interruptions.

- Make sure you know how to get to the interview and arrive in good time.

- Dress and behave appropriately.

- Establish rapport.

- Negotiate a length of time for the interview and stick to it, unless the interviewee is happy to continue.

- Ask open questions, listen to responses and probe where necessary.

- Keep questions short and to the point.

- Avoid jargon, double-barrelled questions and leading questions.

- Listen carefully and acknowledge that you are listening.

- Check recording equipment is working without drawing attention to it.

- Repeat and summarise answers to aid clarity and understanding.

- Achieve closure, thank the interviewee and leave a contact number in case they wish to get in touch with you about anything that has arisen.

- Respect their confidentiality – do not pass on what has been said to third parties unless you have requested permission to do so.

Further reading

Brinkmann, S. and Kvale, S. (2014) *Learning the Craft of Qualitative Research Interviewing*, 3rd edition, Thousand Oaks, CA: Sage.

Keats, D. (2000) *Interviewing: A Practical Guide for Students and Professionals*, Buckingham: Open University Press.

Kvale, S. (2007) *Doing Interviews*, London: Sage.

Rubin, H.J. and Rubin, I.S. (2004) *Qualitative Interviewing: The Art of Hearing Data*, 2nd edition, Thousand Oaks, CA: Sage.

Seidman, I. (2013) *Interviewing as Qualitative Research: A Guide for Researchers in Education and the Social Sciences*, 4th edition, New York, NY: Teachers College Press.

Wengraf, T. (2001) *Qualitative Research Interviewing: Biographic Narrative and Semi-Structured Methods*, Thousand Oaks, CA: Sage.

8
How to Conduct Focus Groups

As we saw in Chapter 3, a focus group is where a number of people are asked to come together in order to discuss a certain issue for the purpose of research. They are popular within the fields of market research, political research and educational research. The focus group is facilitated by a **moderator**, who asks questions, probes for more detail, makes sure the discussion does not digress and tries to ensure that everyone has an input and that no one person dominates the discussion.

If you are interested in running focus groups for your research, you will need to acquire a basic understanding of how people interact in a group setting and learn how to deal with awkward situations (see Table 8).

The role of the moderator

As moderator you must spend some time helping participants to relax. In all focus groups you need to explain the purpose of the group, what is expected of participants and what will happen to the results. Negotiate a length for the discussion and ask that everyone respects, this as it can be very disruptive having people come in late, or leave early. Usually one and a half hours is an ideal length, although some focus groups may last a lot longer.

TIP

The best way to become a successful moderator is through experience and practice. If possible, try to sit in on a focus group run by an experienced moderator. Once you have done this, hold your own pilot focus group, either with friends or actual research participants. You might find it useful to record visually this focus group so that you can assess your body language, see how you deal with awkward situations, analyse how you ask questions, and so on. Don't be disappointed if your first few groups do not go according to plan. Even the most experienced moderators have bad days.

Assure the participants about anonymity and confidentiality, asking also that they respect this and do not pass on what has been said in the group to third parties. You may find it useful to produce and distribute a Code of Ethics (see Chapter 14).

Asking questions

General, easy-to-answer questions should be asked first. Don't expect any type of personal disclosure early in the group. As moderator, listen carefully to everything people say, acknowledging that you are listening by making good eye contact and taking notes regarding issues to which you may return later. Make sure that no one person dominates the discussion, as this will influence your data.

Some moderators prefer to use a list of questions as their interview schedule, whereas others prefer to use a list of topics (see Chapter 7 for more information on developing an interview schedule). The overall aim is a free-flowing discussion within the subject area, and once this happens the input from the moderator may be considerably less than it would be in a one-to-one interview.

Seeking responses

In focus groups you need to try to get as many opinions as possible. You will find that in most focus groups, most people will talk some of the time, although to varying degrees. In some groups, some people may need gentle persuasion to make a contribution. You have to use your discretion about how much you do this, as there may be occasions when somebody is unwilling or too nervous to contribute.

You often find that, even though you have negotiated a time, people enjoy the discussion and want to continue, although at this stage you must make it clear that people can leave, if they wish. Often, some of the most useful and pertinent information is given once the 'official' time is over. Also, you will find that people talk to you on an individual basis after the group has finished, especially those who might have been nervous contributing in a group setting. It is useful to take a notepad and jot down these conversations as soon as possible after the contact, as the information might be relevant to your research.

Finishing the focus group

When you have finished your focus group, thank the participants for taking part and leave a contact name and number in case they wish to follow up any of the issues that have been raised during the discussion. It's good practice to offer a copy of the report to anybody who wants one. However, this might not be practical if the final report is to be an undergraduate dissertation. You could explain this to the participants and hope that they understand, or you could offer to produce a summary report that you can send to them.

Recording equipment

Some market research organisations have purpose-built viewing facilities with one-way mirrors and built-in visual recording equipment. These facilities can be hired at a price which, unfortunately, tends to be beyond the budgets of most students and community groups.

Your local college or university might have a room that can be set up with visual recording equipment and the institution may provide an experienced person to operate the machinery. If your institution doesn't provide this facility, think about whether you actually need a visual recording of your focus group, as the more equipment you use, the more potential there is for things to go wrong. Most social researchers find that an audio recording of the discussion supplemented by a few handwritten notes is adequate (see Chapter 7 for further discussion on different methods of recording).

Using audio recording equipment

Your recorder needs to be powerful enough to pick up every voice. Ideally, it needs to be small and unobtrusive with an inbuilt microphone and a battery indicator light, so that you can check it is still working throughout the discussion, without drawing attention to the machine. If using a tape-recorder, a self-turning facility is useful, as you get twice as much recording without having to turn over the tape. Digital recording equipment tends to have much more space available.

TIP

When recording focus groups I tend to use two recorders, placed in the middle of the group, but at a distance from each other. This means that if one fails I still have a recording from the other and it ensures that all voices in the group can be heard.

TABLE 8: STRATEGIES FOR DEALING WITH AWKWARD SITUATIONS

SITUATION	STRATEGY
Break-away conversations	Say: 'I'm sorry, would you mind rejoining the group, as this is really interesting?'
Digressions	Say: 'That's interesting, what do the rest of you think about . . .' (back to the topic)
Silences	Remain silent. Someone will speak as they will begin to feel uncomfortable. If no one does, ask the question again.
Dominance	First of all, stop making eye contact and look at other people expectantly. If this fails, say: 'Thank you for your contribution. Can we get some opinions from the rest of you please?' Or, 'What do the rest of you think about that?' (This should counteract the one dominant argument by receiving other views on the same issue.)
Leadership	If it is obvious from the start that you have a clear leader who will influence the rest of the group, try to give them another role which takes them away from the discussion, such as handing out refreshments or taking notes. If, however, leadership tendencies aren't immediately obvious, but manifest themselves during the discussion, try to deal with them as with 'dominance', above. If this still fails, as a last resort you might have to be blunt: 'Can you let others express their opinions, as I need to get as wide a variety as possible?' I've actually had to cut short one group and rearrange it when that person wasn't present. The other members were happy to do this, as they were free to express themselves, and their opinions were quite different from those of their self-appointed 'leader'.
Disruption by participants	On rare occasions I have come across individuals who want to disrupt the discussion as much as possible. They will do this in a number of ways, from laughing to getting up and walking around. I try to overcome these from the start by discussing and reaching an agreement on how participants should behave. Usually I will find that if someone does become disruptive, I can ask them to adhere to what we all agreed at the beginning. Sometimes, the other participants will ask them to behave, which often has a greater influence.
Defensiveness	Make sure that nobody has been forced to attend and that they have all come by their own free will. Be empathetic – understand what questions or topics could upset people and make them defensive. Try to avoid these if possible, or leave them until the end of the discussion when people are more relaxed.

The recorder should be placed on a non-vibratory surface at an equal distance from each participant so that every voice can be heard. Before the participants arrive, place it in the centre of the room and test your voice from each seat, varying your pitch and tone. Participants in focus groups tend to speak quietly at the beginning, but once they begin to relax, they tend to raise their voices. Be aware of any noise that could disrupt the recorder, such as ticking clocks or traffic outside.

Choosing a venue

It is extremely important to make sure you choose the right venue for your focus group, as this will affect participation levels, the level of discussion and the standard of recording. You should ask yourself the following questions when considering a venue:

- Is the venue accessible in terms of physical access for those with mobility difficulties?
- Is it accessible in terms of 'mental' access, that is, would the type of people you intend to recruit feel comfortable entering that building?
- Is the building easy to find and the room easy to locate?
- Is free parking available close by?
- Is it accessible by public transport?
- Is the room big enough to accommodate the number of people you intend to recruit?
- Are there enough chairs and are they comfortable?
- Is there anything that could distract the participants (loud noises, telephones, doorbells, people entering the room, people walking past windows, etc.)?
- Is there anything that could disrupt the recording (ticking clocks, drink machines, traffic outside, etc.)?

Once you have chosen your venue, you need to arrive early to make sure that the seating is arranged in an appropriate manner. There is no set rule for this – think about your participants and arrange it accordingly. For example, business people might prefer a boardroom-style seating arrangement, whereas adult learners may prefer an informal seminar-style arrangement.

Recruiting your participants

Without participants you have no focus group. It is essential, therefore, that you take time to ensure that you achieve a high turn-out for your focus group. Think about the following points when recruiting your participants:

- The ideal number of participants is nine or eleven. Odd numbers work better than even numbers, as it is harder for people to pair up in break-away conversations.

- Over-recruit by between three and five people, as some participants, despite reassurances to the contrary, will not attend.

- Offer incentives. If you cannot afford to pay participants, offer refreshments such as wine, soft drinks and nibbles. Do not provide too much wine, as a drunken discussion is not productive, and think about the cultural and/or religious background of participants before using alcohol as an incentive.

- The goal is to achieve a free-flowing, useful and interesting discussion. Think about whether the people you are recruiting would be able to chat to each other in everyday life. People must have some sort of common bond to be able to feel comfortable in a group and this will probably be the focus of your research.

- When someone has agreed to take part in a focus group, contact them a week in advance to let them know about the venue, date and time. Telephone them the day before to make sure they have remembered and are still able to attend.

- Never force, bully or cajole someone into taking part. If someone else is arranging the focus group for you, make sure they do not force or cajole people into participating. Someone who does not want to attend usually makes it clear during the discussion. This can have a detrimental effect on the whole group.

- If someone is in a position of power, they should not be included in the group, as it may stop others airing their opinions, although it is not always possible to prevent such people from attending, as Example 7 illustrates.

EXAMPLE 7: SIMON

I was conducting a focus group with workers in a toy factory. Everyone in the group worked on a production line, but unfortunately their supervisor, who'd arranged for me to run the group, insisted on sitting in on the group. I'd worked in that factory as a temp over the summer holidays and I knew that the supervisor was viewed as a bit of a tyrant. I'd tried desperately to make sure he didn't come to the discussion, but there was nothing I could do.

Sure enough, throughout the discussion if someone said something which was against company policy he would contradict them or say that it simply wasn't true.

In the end people just stopped talking. I had to go and see each person after the group and I got some really interesting information when the supervisor wasn't present. But of course it wasn't recorded properly and I was unsure of how I could use that information in my research. I wanted to arrange another group, but I just knew it would be impossible without the supervisor present. In the end I had to forget about that place and hold another focus group in another factory.

Summary

- Find a suitable venue and check availability. Is it accessible physically and mentally?

- Visit the venue and check it is free from background noises, distractions and interruptions.

- Obtain appropriate recording equipment and practise using it.
- Try your recording equipment in the venue to test suitability.
- Contact participants and check availability for time and place.
- Over-recruit to ensure enough participants.
- Telephone participants the day before the focus group to check they're still intending to participate.
- Arrive at the venue early and arrange the seating in a way that will suit the group.
- Test the recording equipment from each seat.
- Lay out refreshments away from the recorder.
- Greet participants with drinks and nibbles.
- Introduce yourself; explain what the group is about, what is expected of the participants, who the research is for and what will happen to the results.
- Negotiate a discussion length and ask that no one leaves early.
- Discuss issues of confidentiality, anonymity and personal disclosure.
- Start the recorder and begin with general, easy-to-answer questions.
- Watch for group dynamics and deal with them accordingly.
- Listen and take notes.
- Ask questions and probe for more detail.
- Wind up within the negotiated time, unless participants wish to continue.

- Thank participants and give them your name and contact number in case they wish to follow up any of the issues with you.

- Send a summary report to anyone interested.

Further reading

Barbour, R. (2007) *Doing Focus Groups*, London: Sage.

Greenbaum, T.L. (2000) *Moderating Focus Groups: A Practical Guide for Group Facilitation*, Thousand Oaks, CA: Sage.

Krueger, R.A. and Casey, M.A. (2015) *Focus Groups: A Practical Guide for Applied Research*, 5th edition, Thousand Oaks, CA: Sage.

Stewart, D., et al. (2015) *Focus Groups: Theory and Practice*, 3rd edition, Thousand Oaks, CA: Sage.

9
How to Construct Questionnaires

Once you have decided that a questionnaire is the most appropriate data collection method for your research, before you go on to construct the questionnaire you need to think about what, exactly, you want from your research. Too often researchers rush into designing a questionnaire, only to find that it is not yielding the type of information they require.

Chapters 1 and 2 should have helped you to think about your topic and to decide what it is that you actually want from your research. If you are still unsure, talk to your tutor or boss to help you to clarify your thoughts. Time spent getting this right now will save much wasted time as your questionnaire design progresses.

Deciding which questionnaire to use

If you're sure that a questionnaire is the most appropriate method for your research, you need to decide whether you intend to construct a closed-ended, open-ended or combination questionnaire. In open questions, respondents use their own words to answer a question, whereas in closed questions prewritten response categories are provided.

The advantages and disadvantages of open and closed questionnaires are discussed in Table 9. You need to think about whether your questionnaire is to be self-administered – that is, the respondent fills it in on his own, away from the researcher – or whether it is to be interviewer administered. Self-administered questionnaires could be sent through the post, delivered in person or distributed online. It is also important to think about the analysis of your questionnaire at this stage, as this could influence its design (see Chapter 12).

TABLE 9:
OPEN AND CLOSED QUESTIONS: ADVANTAGES AND DISADVANTAGES

OPEN QUESTIONS	CLOSED QUESTIONS
Tend to be slower to administer.	Tend to be quicker to administer.
Can be harder to record responses.	Often easier and quicker for the researcher to record responses.
May be difficult to code, especially if multiple answers are given.	Tend to be easy to code.
Do not stifle response.	Respondents can only answer in a predefined way.
Enable respondents to raise new issues.	New issues cannot be raised.
Respondents tend to feel that they have been able to speak their mind.	Respondents can only answer in a way that may not match their actual opinion and may, therefore, become frustrated.
In self-administered questionnaires, respondents might not be willing to write a long answer and may decide to leave the question blank. How do you know the meaning of a blank answer when you come to the analysis?	Is quick and easy for respondents to tick boxes – might be more likely to answer all the questions.
Can be used to find out all the possible responses before designing a closed-ended questionnaire.	Can include a section at the end of the questionnaire for people to write in a longer response if they wish.

Wording and structure of questions

When constructing each question think about the words you use. Avoid jargon and technical terms whenever possible. Try not to use words that may have a double meaning or be misinterpreted, as some words have different meanings for different groups of people. Don't use emotive words. Make sure the question is not ambiguous. And, above all, avoid questions that will cause annoyance, frustration, offence, embarrassment or sadness. You should never make someone feel uncomfortable, for whatever reason, as a result of filling in your questionnaire.

- **Questions should be kept short and simple.** This will avoid many of the problems outlined above. Check that a question is not double-barrelled – that is, two questions in one. If it is, ask two questions rather than one. Also, avoid negative questions – the type that have 'not' in them – as this can be confusing, especially when a respondent is asked to agree or disagree.

- **Make sure that your questions don't contain some type of *prestige bias*.** This phrase refers to questions that could embarrass or force respondents into giving a false answer. They might do this if they do not want to look 'bad' in front of the researcher, or because it is expected behaviour. Questions about income or educational qualifications might elicit this type of response, so you need to be careful about how you try to obtain this information.

- **Some issues may be very sensitive and you might be better asking an indirect question rather than a direct one.** Promising confidentiality and anonymity may help, but many respondents can, understandably, be sceptical about these promises. If you ask an indirect question in which respondents can relate their answer to other people, they may be more willing to answer the question.

Using closed-ended questions

If you are constructing a closed-ended question, try to make sure that all possible answers are covered. This is particularly important for time and frequency questions such as 'How often do you . . .' You need to make sure that all the frequencies are covered so that respondents aren't constrained in their answers and tick a box that isn't right for them. Also, you want to make sure that you don't create opinions artificially by asking someone a question about which they don't know, or don't care. You need to make sure that you include a 'don't know' category in this case.

EXAMPLE

My partner is self-employed and works on his own in an office based at home. He was recently asked to fill in a questionnaire that included the following question

Do your work colleagues swear . . .

More than you? ☐

About the same as you? ☐

Less than you? ☐

They don t swear. ☐

My partner does not have work colleagues. This not only made the question irrelevant to his work situation, but also made him feel that the whole questionnaire was irrelevant and, therefore, he threw the questionnaire into the bin.

If you expect people to take the time to fill in your questionnaire, you must make sure that they can answer the questions and that they feel they are relevant and worthwhile. Not only does this question presume that the respondent has work colleagues, but it also presumes that the respondent swears, which can cause offence to some people.

Avoiding leading questions

Don't asking leading questions. The question 'How often do you wash your car?' might seem innocuous enough. However, it makes two assumptions. First, it assumes that the respondent has a car and second, it assumes the respondent washes his car. It could be considered a prestige bias question. Would a respondent feel bad if they didn't have a car and therefore tick 'four times a week' anyway? Would they feel bad if they don't ever wash their car but feel the researcher expects them to? If you need to ask this question, you should ask a filter question first to find out whether the respondent actually owns a car. Then you would need to ask: 'If you wash your car, how many times a year do you do so?' By wording the question in this way and by being careful about the frequency list, you're not leading the respondent into answering in a certain way.

Have a look at Exercise 2, which will help you to think about some of the issues involved in the wording and structuring of questions.

EXERCISE 2

Read the following questions and decide what is wrong with them. Think about how you might overcome the problems you have identified.

1. Do you go swimming?

Never	☐
Rarely	☐
Frequently	☐
Sometimes	☐

2. What do you think about the Green Peace attempt to blackmail the Government?

3. What is wrong with the young people of today and what can we do about it?

4. How many books have you read in the last year?

None	☐
1–10	☐
10–20	☐
20–30	☐
Over 30	☐

5. What is the profit of your company, to the nearest one hundred pounds?

6. What do you think should be done about global warming?

Points to consider

1. The problem with this question is in the categories supplied for the answer. Everybody has a different idea as to what words such as 'sometimes' and 'frequently' mean. Instead, give specific time frames such as 'twice a year' or 'once a month'. Also, the order of answers should follow a logical sequence – in the example above, they do not.

2. This is a very leading question which uses an emotive word – blackmail. It assumes that Green Peace is blackmailing the Government and that someone knows about the issues and would be able to answer. A filter question would have to be used in this case and the word 'blackmail' changed.

3. This question is double-barrelled, leading and ambiguous. It asks two questions in one and so needs to be split up. The word 'wrong' is emotive and suggests there is something not normal about the young people of today. It asks the respondent to distance themselves and comment from the moral high ground.

4. This question may contain prestige bias – would people be more likely to say they have read plenty of books when they might not have read any? Also, the categories for the answers need modification – which box should someone who answered '20' tick?

5. This question assumes knowledge and could only be asked of someone who has the figures to hand. It also asks for what could be confidential information that a respondent might be reluctant to give. The word 'profit' has different meanings for different people, especially if the question is asked by an interviewer, rather than read by the respondent. In an often quoted case, when this question was posed to them verbally, the respondent took it to mean 'prophet' and as such was unable to answer the question.

6. In this question it is assumed that the respondent thinks something should be done about global warming and that they are able to comment on the issue. Indeed, global warming in itself is a contested issue. The question leads the respondent into having an opinion about something on which they might not otherwise have one.

Length and ordering of questions

When you're constructing a questionnaire, keep it as short as possible. If it has to be longer because of the nature of your research, think about whether your respondents will actually take the time to fill it in. Some people will do so if they feel there is a personal benefit to be gained. This is why long consumer behaviour surveys offer entry into large prize draws for completed questionnaires. If your budget is limited, you might be able to offer a copy of the final report or other information that may be of use to the respondent as an incentive.

Be realistic about how long a questionnaire will take to fill in. Saying it will take a 'moment' is not helpful. Include filter questions with answers such as 'If no, go to question 28'. Psychologically it's good for respondents to be able to jump sections, as it stops them becoming frustrated by unnecessary or irrelevant questions.

As with interviewing or focus groups, when designing a questionnaire start with easy questions that respondents will enjoy answering, thus encouraging them to continue filling in the questionnaire. If you begin with complex questions that need long responses, your respondents will be less likely to fill in the form. If you're constructing a combined questionnaire, keep your open-ended questions for the end as, once someone has spent time completing the rest of the questionnaire, they are more likely to continue with those questions that take a little more effort to complete.

When constructing a questionnaire, you need to make it as interesting as possible and easy to follow. Try to vary the type and length of questions, as variety provides interest. Group the questions into specific topics, as this makes it easier to understand and follow. Layout and spacing are

TIP

When deciding how long your questionnaire should be, think about the topic and the type of people who you are hoping will fill it in. People are more likely to spend longer answering questions about a topic that they feel strongly about, or with which they are very involved. They are less likely to spend a long time answering questions about a topic that does not interest them or that they view as unimportant or frivolous.

extremely important. If your questionnaire looks cluttered, respondents will be less likely to fill it in.

Collecting personal information

Opinion is divided about where personal information should be included on a questionnaire. I tend to include it at the end, as I believe people are more likely to fill in this information when they have already invested time and energy in completing the rest of the form.

As the researcher, you need to think about collecting only that type of personal information that is completely relevant to your research. Be sensitive to the type of information people will be reluctant to give. An example of a personal profile section is provided in Fig. 1. This form may seem short, but it is important not to ask for too much personal information as respondents will become suspicious and want to know why you want the information. You also need to assure them that you understand and will comply with the Data Protection Act (see Chapter 14). This Act updates and modernises data protection laws in the UK. It also implements the European Union's General Data Protection Regulation (GDPR) in national law, ready for withdrawal of the UK from the Union in 2019 (see Chapter 14).

Piloting the questionnaire

Once you have constructed your questionnaire, you must pilot it. This means that you must test it out to see if it is obtaining the results you require.

First of all, ask people who have not been involved in its construction to read it through and see if there are any ambiguities that you have not noticed.

Once this has been done, alter the questions accordingly, then send out a number of questionnaires to the type of people who will be taking part in the main survey. Make sure they know it is a pilot test and ask them to forward any comments they may have about the length, structure and wording of the questionnaire. Go through each response very carefully, noting comments and looking at the answers to the questions, as this will help you to discover whether there are still ambiguities present.

1. Are you:

Female? ☐ Male? ☐

2. What is your age?

Under 26 ☐ 26–35 ☐ 36–45 ☐
46–55 ☐ 56–65 ☐ Over 65 ☐

3. To which of these ethnic groups do you consider you belong?
(These categories were used in the 2001 Census and have been reproduced for ease of comparison.)

White:

British ☐ Any other White background (please describe) ☐

Mixed:

White and Black Caribbean ☐ White and Black African ☐
White and Asian ☐ Any other Mixed background (please describe) ☐

Asian or Asian British:

Indian ☐ Pakistani ☐ Bangladeshi ☐
Any other Asian background (please describe) ☐

Black or Black British

Caribbean ☐ African ☐
Any other Black background (please describe) ☐

Chinese or other Ethnic Group

Chinese ☐ Any other (please describe) ☐

Which of the following categories apply to you? (Please tick all that apply.)

Housewife/husband ☐ Full-time student ☐
Part-time student ☐ Self-employed ☐
Seeking paid employment ☐ In part-time, paid employment ☐
In full-time, paid employment ☐

Fig. 1. Personal profile form

Alter the questionnaire again. If you have had to undertake major alterations, you may need to pilot the questionnaire again. This may seem a rather long and laborious process, but it is incredibly important, especially if you're intending to send out a large number of questionnaires.

Obtaining a high response

Questionnaires are big business and as more and more fall through our letterboxes we become less willing to spend the time completing them. You need to make yours stand out so that all your careful planning and construction is not wasted. There are many simple measures you can take to try to ensure a high response rate.

- Is the questionnaire relevant to the lives, attitudes and beliefs of the respondents?
- Can the respondents read if they are to be given a self-administered questionnaire?
- Are there any language issues? Do you need to translate the questionnaire into another language?
- Are your intended respondents likely to cooperate? For example, illegal immigrants may be less likely to fill in a questionnaire than legal immigrants.
- Is the questionnaire well constructed and well laid out?
- Is it clear, concise and uncluttered?
- Are the instructions straightforward and realistic about how long it will take to complete?
- Has the respondent been told who the research is for and what will happen to the results?
- Has the respondent been reassured that you understand and will comply with the Data Protection Act?

- Has the questionnaire been piloted to iron out any problems?
- Can your respondents see some personal benefit to be gained by completing the questionnaire?
- Is return postage included?
- Has a follow-up letter and duplicate questionnaire been sent in cases of non response?

QUESTIONNAIRE DESIGN CHECKLIST

- Make your questionnaire as short as possible.
- Make sure people will be able to answer your questions.
- Don't assume knowledge or make it seem that you expect a certain level of knowledge by the way your questions are worded.
- Start with easy-to-answer questions. Keep complex questions for the end.
- Ask for personal information at the end.
- Use a mix of question formats.
- Don't cause offence, frustration, sadness or anger.
- Avoid double-barrelled questions.
- Avoid words with emotional connotations.
- Avoid negative questions.
- Avoid jargon and technical words.
- Avoid words with multiple meanings.

- Avoid leading questions.
- Avoid vague words such as 'often' and 'sometimes'.
- Provide all possible responses in a closed question.
- Consider as many alternatives as possible.
- Use specific time frames when asking about behaviour.
- Use specific place frames, e.g. 'In which country were you born?'

Summary

- Think about how you're going to analyse your survey prior to constructing your questionnaire.
- Decide whether you're interested in behaviour, beliefs, attitudes or characteristics, or a combination of the above.
- Make sure you have made the right decisions concerning open-ended questions, closed-ended questions or a combination of both.
- Decide whether your questionnaire is to be self-administered or interviewer administered.
- Think about how you intend to distribute your questionnaire, e.g. by hand, through the post or via the internet.
- Construct the questionnaire adhering to the checklist produced above.
- Include a covering letter with information about who the research is for and what will happen to the results.
- Include instructions on how to complete the questionnaire.
- Include details about how the questionnaire is to be returned (making sure

you enclose a pre-paid envelope if you need the respondent to return the form to you).

- Make sure you include a date by which time you would like the questionnaire returned.
- Pilot the questionnaire and instructions to check that all can be understood.
- Amend accordingly and re-pilot.
- Send out/administer questionnaire.
- Send out follow-up letters and questionnaires to non-responders.

Further reading

Aldridge, A. and Levine, K. (2001) *Surveying the Social World: Principles and Practice in Survey Research*, Buckingham: Open University Press.

Bradburn, N. (2004) *Asking Questions: The Definitive Guide to Questionnaire Design*, San Francisco, CA: John Wiley and Sons, Inc.

Cox, J. and Cox, K. (2008) *Your Opinion Please: How to Build the Best Questionnaire in the Field of Education*, Thousand Oaks, CA: Corwin Press.

Lance, C.E. and Vandenberg, R.J. (eds) (2015) *More Statistical and Methodological Myths and Urban Legends*, New York, NY: Routledge.

Tourangeau, R., Rips, L.J. and Rasinski, K. (2000) *The Psychology of Survey Response*, Cambridge: Cambridge University Press.

10
How to Carry Out Participant Observation

Participant observation emerged as the principal approach to ethnographic research, which seeks to provide descriptive studies of human societies (see Chapter 2).

Participant observation can be viewed as a methodology, rather than a method, as it took shape within particular historical and social circumstances within anthropology and sociology. It is a procedure for generating understanding about the way of life of others. However, as there are many practical 'how to' issues involved in the use of participant observation, I am going to discuss it as a research method. But as you will see, there are several methodological issues that are raised in the following discussion, especially concerning ethics and the personal role of the researcher.

Places of study

Participant observation can be carried out within any community, culture or context that is different to the usual community and/or culture of the researcher. It may be carried out within a remote African tribe or in hospitals, factories, schools, prisons and so on, within your own country. The researcher immerses herself into the community – the action is deliberate and intended to add to knowledge.

The researcher participates in the community while observing others within that community, and as such she must be a researcher 24 hours a day. In practice most researchers find that they play more of a role as an observer than they do as a participant.

Gaining access

Participant observation, as a research method, cannot work unless you're able to gain access to the community that you wish to study. Before you spend a lot of time planning your project, you need to find out whether you can obtain this access. The level of negotiation required will depend upon the community, culture or context. If it is a culture with which you already have a certain amount of familiarity, and vice versa, you should find it easier to gain access. However, if it is a secret or suspicious community, you may find it much harder to gain access.

Using gatekeepers

If you do expect to encounter difficulties, one way to overcome this problem is to befriend a member of that community who could act as a gatekeeper and help you to get to know other people. Obviously, it is important to spend time building up the required level of trust before you can expect someone to introduce you into their community.

If it is not possible to befriend a member of the community, you may have to approach the person or committee in charge, first by letter and then in person. You have to be persuasive.

Creating a good impression

First impressions are important and you need to make sure that you dress and act appropriately within the community. You must not appear threatening in any way. Some people will be suspicious of the motives of a researcher, especially if they're not familiar with the research process. In the early stages it is better to answer any questions or suspicions directly and honestly, rather than try to avoid them or shrug them off.

Acting ethically

Because of the nature of participant observation, there tend to be more issues involving ethics and morals to consider. As you intend to become part of a specific group, will you be expected to undertake anything illegal? This could happen with research into drug use or crime syndicates, where people may not trust you until you become one of them and join in their activities. Would you be prepared to do this and put up with any consequences that could arise as a result of your activities?

If the group is suspicious, do you intend to be completely honest about who you are and what you're doing? Are you prepared to lie if it means you can gain access? How would you deal with any problems that may arise as a consequence of your deception?

What if your participation within a group causes problems, anxiety or argument among other members? Would you know how to deal with the situation? Would you be prepared to withdraw and ruin all your hard work for the sake of your informants? These issues and others are covered in more depth in Chapter 14. Also, there are many personal considerations and dilemmas that you need to think about before undertaking participant observation, as illustrated below.

TIP

If you find that your presence initially causes a sense of personal discomfort among some participants, it is important to record and analyse this discomfort. This is because it can provide deeper insight into the personal behaviour, actions and motivation of people involved in your research. Similarly, it is important to record and analyse other reactions to your presence, such as an over-eagerness to please or a constant challenging of your words.

PERSONAL CONSIDERATIONS WHEN ENTERING THE FIELD

Some people will not accept you. Are you prepared for rejection and can you handle it when it happens? Are you able to banish rejection anxieties from the outset?

Will your contact be traumatic or upsetting? If so, can you handle this?

If you're going to come across people with very different social and political beliefs, can you remain neutral and keep your opinions to yourself? Some researchers may try arguing their point in the hope that they will get more information and it will deepen their understanding. However, you must be careful not to compromise your position.

Are you prepared for the emergence of as yet unconscious emotional factors? You may find out things about yourself that you do not like, especially in terms of your own prejudices.

Are you prepared to be used as a scapegoat if things go wrong within the community under study? Often it is easier for the community to blame an outsider and many researchers are happy to go along with this because they know they will be leaving the community at some point.

Can you handle the feelings of guilt that may arise as a result of the roles you've got to play?

Are you prepared to pretend to have feelings that may not be real? Some researchers would argue that you should not do this because it is being dishonest. The level to which you are prepared to do this has to be your own choice.

Are you aware of your own likes and dislikes? Can you put stereotypes and prejudices aside?

Collecting and analysing information

At the beginning of a participant observation study it is hard to understand what everything means. At first something may appear very significant, but later it might become a minor detail. However, you will not know this until you have started to analyse what is going on. So it is important from the outset to start taking copious notes. You need to have a good memory, as in many situations it is not possible to take notes at the time. You need to have a notepad, laptop or mobile device (and sufficient battery life) with you at all times that will enable you to record your thoughts during, or soon after, the encounter.

All researchers ask questions. However, in the beginning stages of a participant observation study, it is better to seek information by not asking questions. This is because it is hard for you to know what to ask at this stage. Instead, you will find that people come to you and ask questions. This in itself is valuable information and can tell you a lot about those people, so all questions should be noted and analysed.

Field notes

Field notes are your main way of recording data. These might be practical details about events, times, dates and places. Or they might be methodological notes concerning your role, your influence on the encounter, your relationship with the informants, sampling procedures and so on. As time moves on, your notes will be to do with a preliminary analysis and the forming of hypotheses, which you can go on to check out with your informants. They may be observations on what a specific encounter might mean. Also, as your research progresses you will start to code and classify your notes (see Chapter 12).

Taking notes is a very personal process and you need to find a method that will suit you. Many researchers develop their own form of shorthand, but if you do this keep it simple because, if your contact is over a long period of time, you may not understand the shorthand you used at the beginning. Most researchers keep a day-to-day diary in addition to all the other field notes. You will also need to keep all transcripts of interviews, photographs, maps, tapes, visual recordings, diagrams and plans. Everything needs to be recorded and stored systematically, so good organisational skills are important if you wish to undertake participant observation.

IT equipment

Some researchers find it useful to obtain a laptop for recording and storing information when in the field, but this will depend on the situation. Indeed, this applies to all IT equipment – you will need to make sure that there is safe and secure storage and that you have obtained appropriate insurance, as well as checking that your participants are happy with you recording and storing information in this way.

Data analysis

Most of your analysis takes place in the field so that you can cross check and verify your hypotheses. At this stage you will find that you'll have a number of key informants who will be able to help you with this process. This is very useful as they'll tell you if you're on the wrong track.

Withdrawing from the field

When you have been immersed in a particular culture for a long period of time, it can be hard to break away. Indeed, some researchers have found that they do not want to do so, although this only happens rarely. If, however, you have remained connected to your role as researcher, you will know when it is time to break away, write up your results and pass on what you have learnt.

It is important to leave your community on good terms. Many researchers find that it is helpful to stay in touch with their contacts – these people will want to see what is written about them. They will be interested and may still have comments to make. You may also wish to return to your community several years later and conduct a follow-up study.

Finally, you must make sure that you try not to do anything that will give researchers a bad name and cause problems for other researchers who may wish to follow in your footsteps (see Chapter 14).

Summary

- In participant observation, the researcher immerses herself into a community, culture or context. The action is deliberate and intended to add to knowledge.

- A participant observer is a researcher 24 hours a day.

- To gain access a researcher must be non-threatening, displaying appropriate behaviour and body language and wearing appropriate dress.

- A useful way of gaining access is to find a gatekeeper who can introduce you to other members of the community.

- A researcher needs to do much soul-searching before going into the field, as the experience can raise many ethical, moral and personal dilemmas.

- It is sometimes quicker and more economical to wait for questions to come to the researcher, rather than ask questions of informants in the early stages of a study.

- Field notes may record practical details, methodological issues, personal thoughts, preliminary analyses and working hypotheses.

- Data analysis takes place in the field so that hypotheses can be discussed with key informants.

- The community should be left on good terms and any written reports should be given back to the people for their interest and personal comments.

Further reading

Angrosino, M. (2007) *Doing Ethnographic and Observational Research,* London: Sage.
DeWalt, K. and DeWalt, B. (2011) *Participant Observation: A Guide for Fieldworkers,* 2nd edition, Lanham, MD: AltaMira Press.
Gobo, G. (2004) *Doing Ethnography,* London: Sage.
Whyte, W.F. (1997) *Creative Problem-Solving in the Field: Reflections on a Career,* Walnut Creek: Alta Mira.

11
How To Conduct Experiments

As we have seen in Chapter 3, there are three main types of experiment involving human beings that are conducted by social researchers: controlled experiments, field experiments and natural experiments. They are used in subjects such as politics, education, economics, marketing, medicine, health, business, sociology and psychology.

This type of research seeks to add to knowledge through careful inquiry that involves systematic and controlled testing, scientific observation and logical analysis to understand what is happening. It moves through a series of stages (often involving repetition) developing and expanding on previous research and moving through different sequences to explain emerging ideas and significant findings.

Understanding the scientific method

This type of experimental research is commonly referred to as the 'scientific method'. This is a method of investigation based on prediction, experiment and observation to understand causal processes. If you are interested in this type of research it is important that you get to grips with the rules, procedures and techniques of the scientific method so that you can approach your experiment in an objective, standardised way to improve your results. This enables you to consider all the evidence before you make a statement of fact.

Some researchers believe that this is the only way to conduct 'proper' research because it is objective and eliminates bias. Others, however, believe that it is impossible to be completely objective and eliminate bias, even if we follow correct procedures and techniques, because we are all human beings that are influenced by personal and structural bias, culture and society. In this case, bias cannot be eliminated but can only be acknowledged, controlled and/or reduced.

Developing and running experiments

The procedure for running an experiment may vary slightly, depending on the type of experiment you intend to conduct. However, you will need to undertake the following for all types of experiment (the order, length and type of task can vary, depending on whether you are interested in controlled, field or natural experiments):

1. Make an observation that leads to a question that could be, or needs to be, investigated.

2. Develop a title that describes the nature of the experiment in a clear and concise way.

3. Produce a statement of purpose that is clear and succinct. This should describe what you want to learn from your experiment and will help you to focus your thoughts on the specific point of investigation.

4. Describe the methods that you intend to use. How will you conduct your experiment? What are the different stages and what methods will you use for each stage?

5. Gather information about the topic. What other research has covered this topic? What literature, resources and background information are available?

6. Identify and summarise the main points from the existing literature.

7. Formulate your hypothesis.

8. Design your experiment, paying attention to the supply and acquisition of materials, and health and safety concerns.

9. Identify participants and describe recruitment processes, paying attention to ethical issues such as informed consent, anonymity, confidentiality and privacy.

10. Develop a data management and sharing plan, to include issues of data security and data protection.

11. Apply for, and await, ethical approval.

12. Choose and/or recruit participants.

13. Run your experiment. This involves changing one or more variable in each experiment, observing, recording and measuring results, followed by repetition to verify findings.

14. Summarise results.

15. Draw conclusions.

16. Write up your report.

17. Publish findings (produce a dissertation, thesis, journal paper, website, blog or make a presentation, for example).

TIP

Note that experiments can be creative, dynamic, exciting and unpredictable. Conclusions can be reached, modified, revised and changed. Your experiment could raise more questions than it answers.

Avoiding mistakes in experiments

There are a number of mistakes that can be made when you run your experiment. An awareness of these, and the action that you can take to avoid these mistakes, will help you to conduct successful experiments. These are listed in Table 10.

Ensuring validity and reliability

When conducting experiments, close attention must be paid to issues of validity and reliability. These are the ways that other researchers, colleagues, tutors and examiners judge the value, importance, correctness and generalisability of your work.

Validity

Validity refers to the accuracy of the measurement, asking whether the tests that have been used by the researcher are measuring what they are supposed to measure. Your work must be shown to be sound logically and factually so that the results can be generalised to other people, other settings and over time.

TABLE 10: RUNNING EXPERIMENTS: OVERCOMING MISTAKES AND FINDING SOLUTIONS

MISTAKE	SOLUTION
The experiment is unfocused.	Produce a title that describes the nature of the experiment in a clear and concise way. Try to sum up, in one sentence, your research. Discuss the sentence with your tutor, boss or your peers.
Test and purpose don't match.	Develop a statement of purpose, which describes clearly what you intend to do in your experiment. Discuss with your tutor, boss or supervisor.
Incorrect procedures are used.	Take time to choose appropriate methods and ensure that they fit the stated purpose.
Researcher bias is introduced.	Find out what is meant by researcher bias. Be aware of the problems that can occur, so that they can be avoided. This could include selection bias, design bias, measurement bias and reporting bias, for example.
Hypothesis is presumed to be right.	Don't presume that the hypothesis is right, even if it seems obvious or common sense. Test thoroughly with an open mind.
Data are ignored.	Remain observant and report all results, whether or not they support your hypothesis. Maintain high ethical standards.
Data are manipulated.	Find and observe a scientific code of conduct and maintain high ethical standards. Gain more experience and knowledge: read, take courses, seek advice.
Important errors are missed.	Improve observation skills and keep comprehensive notes about everything you observe. Don't get distracted. Ensure that you observe with an open mind and don't be ruled by expectations or personal bias.
Data are interpreted incorrectly.	Become more experienced in data analysis techniques. Attend a course, read around the subject or speak to your tutor.
Falsification of results.	Maintain high moral and ethical standards. Find out about, and adhere to, a code of conduct for research integrity.

Reliability

Reliability refers to the way that the research method is able to yield the same results in repeated trials. It refers to consistency of measurement and asks whether other researchers would get the same results under the same conditions. As we have seen in Chapter 3, some types of experiment are harder to replicate than others. However, by paying close attention to correct procedures and through careful and systematic analysis and reporting, you will increase reliability and make it easier for others to replicate your work.

There are different types of validity and reliability. If you are interested in conducting experiments, you must find out about these. The books listed at the end of this chapter give more detail on these issues.

> **TIP**
>
> Pay careful attention to detail and develop your understanding of experimental processes and statistical techniques, as this will help you to undertake valid and reliable research. Read around the subject, seek further advice and enrol on relevant courses.

Experimenting ethically

It is important that you think about ethical issues and ensure that you conduct your experiment in an ethical way if you intend to carry out any type of experiment involving human beings for your research. You also need to obtain ethical approval (see Chapter 1). Questions that you need to ask include:

- Will taking part in the study cause harm to my participants? This could be physical, psychological or financial harm, for example.
- What rights and welfare issues should I consider?
- How am I going to ensure that I treat participants with respect?
- Do I need to obtain informed consent? If so, how am I going to do this? How can I ensure that this consent is given voluntarily and freely by a person who is able to make the decision?
- How am I going to ensure that I don't mislead participants or try to persuade them to take part when it might not be the best course of action for them?

- Am I in a position of authority that might pressurise people to take part?

- Do I understand what permissions are required and know how to go about gaining them (ethical approval, organisation permission and parental consent, for example).

- How will I ensure anonymity, confidentiality and privacy?

- How am I going to store and manage data safely and securely?

- Am I going to share data and how do I make participants aware of this?

- Am I fully aware of my legal obligations (for example, issues of child protection and appropriate police check requirements)?

- Has a risk assessment been carried out to identify things that could go wrong, make contingency plans and identify risks that can be avoided?

- Are there any conflicts of interest in my research? If there are, what action do I need to take?

TIP

Note that experiments have an influence *on* society and culture (new discoveries and development, and the benefits they bring) and are influenced *by* society and culture (the type of experiments conducted, the subjects covered and the methods used, for example).

Summary

- Experimental research is a method of investigation based on prediction, testing and observation.

- There are three main types of experiment used by social researchers: controlled experiments, field experiments and natural experiments.

- You must pay attention to issues of bias, including selection bias, design bias, measurement bias and reporting bias.

- When testing your hypothesis you must ensure that you do not ignore, manipulate or change data to fit your hypothesis.

- You must adhere to a strict code of conduct and work with accuracy and integrity when conducting experiments.

- You must pay close attention to issues of validity and reliability.

- You must ensure that all research is carried out in an ethical way: your participants must not suffer harm and informed consent must be obtained. Ethical approval must be obtained for all research involving human subjects.

Further reading

Blasius, J. and Thiessen, V. (2012) *Assessing the Quality of Survey Data*, London: Sage.

Chang, M. (2014) *Principles of Scientific Methods*, Boca Raton, FL: Taylor & Francis Group.

Duflo, E. and Banerjee, A. (eds) (2017) *Handbook of Field Experiments*, Amsterdam: North-Holland.

Field, A. and Hole, G. (2003) *How to Design and Report Experiments*, London: Sage.

Robinson Kurpius, S. and Stafford, M. (2006) *Testing and Measurement: A User-Friendly Guide*, Thousand Oaks, CA: Sage.

Viceisza, A. (2012) *Treating the Field as a Lab: A Basic Guide to Conducting Economics Experiments for Policymaking*, Washington, DC: International Food Policy Research Institute.

Viswanathan, M. (2005) *Measurement Error and Research Design*, Thousand Oaks, CA: Sage.

Webster, M. and Murray, J. (eds) (2014) *Laboratory Experiments in the Social Sciences*, London: Academic Press.

12
How to Analyse Your Data

The methods you use to analyse your data will depend on whether you have chosen to conduct **qualitative** or **quantitative** research, and this choice will be influenced by personal and methodological preference and educational background. It could be influenced also by the methodological standpoint of the person who teaches on your research methods course.

Deciding which approach to use

For quantitative data analysis, issues of validity and reliability are important (see Chapter 11). Quantitative researchers endeavour to show that their chosen methods succeed in measuring what they purport to measure. They want to make sure that their measurements are stable and consistent and that there are no errors or bias present, either from the respondents or from the researcher.

Qualitative researchers, on the other hand, might acknowledge that participants are influenced by taking part in the research process. They might also acknowledge that researchers bring their own preferences and experience to the project. Qualitative data analysis is a very personal process. Ask two researchers to analyse a transcript and they will probably come up with very different results. This may be because they have studied different subjects, or because they come from different political or methodological standpoints. It is for this reason that some researchers criticise qualitative methods as 'unscientific' or 'unreliable'. This is often because people who come from quantitative backgrounds try to ascribe their methods and processes to qualitative research. This is a fruitless exercise. The two approaches are very different and should be treated as such.

When to analyse data

Quantitative and qualitative data are analysed in different ways. For qualitative data, the researcher might analyse as the research progresses, continually refining and reorganising in light of the emerg ing results.

For quantitative data, the analysis can be left until the end of the data collection process, and if it is a large survey, statistical software is the easiest and most efficient method to use. For this type of analysis time has to be put aside for the data input process, which can be long and laborious, unless your questionnaires can be scanned. However, once this has been done the analysis is quick and efficient, with most software packages producing well-presented graphs, pie charts and tables that can be used for the final report.

Analysing qualitative data

To help you with the analysis of qualitative data, it is useful to produce an **interview summary form** or a **focus group summary form** that you complete as soon as possible after each interview or focus group has taken place. This includes practical details about the time and place, the participants, the duration of the interview or focus group, and details about the content and emerging themes (see Figures 2 and 3). It is useful to complete these forms as soon as possible after the interview and attach them to your transcripts. The forms help to remind you about the contact and are useful when you come to analyse the data.

There are many different types of qualitative data analysis. The method you use will depend on your research topic, your personal preferences and the time, equipment and finances available to you. Also, qualitative data analysis is a very personal process, with few rigid rules and procedures. It is for this reason that each type of analysis is best illustrated through examples (see Examples 8–11 below).

Formats for analysis

To be able to analyse your data you must first of all produce it in a format that can be easily analysed. This might be a transcript from an interview or focus group, a series of written answers on an open-ended questionnaire, or field notes or memos written by the researcher. It is useful to write memos and notes as soon as you begin to collect data, as these help to focus your mind and alert you to significant points that may be coming from the data. These memos and notes can be analysed along with your transcripts or questionnaires.

The qualitative continuum

It is useful to think of the different types of qualitative data analysis as positioned on a continuum (see Figure 4). At the one end are the highly qualitative, reflective types of analysis, whereas on the other end are those that treat the qualitative data in a quantitative way, by counting and coding data.

Interviewee: ____________________ Date of Interview: ____________________
Place: ____________________ Time of Interview: ____________________
Duration of Interview: ____________________

Where did the interview take place? Was the venue suitable? Does anything need to be changed for future interviews?

How easy was it to establish rapport? Were there any problems and how can this be improved for next time?

Did the interview schedule work well? Does it need to be altered or improved?

What were the main themes that arose in the interview? Did any issues arise that need to be added to the interview schedule for next time?

Is the interviewee willing to be contacted again? Have I promised to send any information or supply them with the results or a copy of the transcript?

Fig. 2. Interview summary form

Date: ______________________ Time: ______________________

Venue: ______________________ Duration: ______________________

Group: ______________________

Diagram of seating plan with participant codes:

Where did the focus group take place? Was the venue suitable? Does anything need to be changed for future focus groups?

How many people took part and who were they? Did they work well as a group or were there any adverse group dynamics? What can I learn from this for the next group?

Did the interview schedule work well? Does it need to be altered or improved?

What were the main themes that arose during the focus group? Does anything need to be added to the interview schedule for the next focus group?

Are any of the participants willing to be contacted again? Have I promised to send any further information or the final report to anyone?

Fig. 3. Focus group summary form

Highly Qualitative	Combination	Almost Quantitative
e.g. thematic and comparative analysis	e.g. discourse and conversational analysis	e.g. content analysis
reflexive and intuitive takes place throughout data collection	uses a combination of reflexivity and counting	code and count mechanical task can be left until end of data collection

Fig. 4. Qualitative data analysis continuum

For those at the highly qualitative end of the continuum, data analysis tends to be an ongoing process, taking place throughout the data collection process. The researcher thinks about and reflects upon the emerging themes, adapting and changing the methods if required. For example, a researcher might conduct three interviews using an interview schedule she has developed beforehand. However, during the three interviews she finds that the participants are raising issues that she has not thought about previously. So she refines her interview schedule to include these issues for the next few interviews. This is data analysis. She has thought about what has been said, analysed the words and refined her schedule accordingly.

Thematic analysis

When data is analysed by theme, it is called **thematic analysis**. This type of analysis is highly **inductive**, that is, the themes emerge from the data and are not imposed upon it by the researcher. In this type of analysis, the data collection and analysis take place simultaneously. Even background reading can form part of the analysis process, especially if it can help to explain an emerging theme. This process is illustrated in Example 8.

EXAMPLE 8: RICHARD

Richard was interested in finding out what members of the public thought about higher education. During a focus group with some library workers, he noticed that some people had very clear ideas about higher education, whereas others had very little idea. This was immediate, on-the-spot analysis.

He asked the group why they thought this was the case and it emerged that the people who had clear ideas about higher education had either been to college or university themselves, or knew someone close to them who had been through higher education. This theme had emerged from one group. Richard decided to follow it up by interviewing people who had never been to college or university to see how different their perceptions might be.

Comparative analysis

Closely connected to thematic analysis is **comparative analysis**. Using this method, data from different people is compared and contrasted and the process continues until the researcher is satisfied that no new issues are arising. Comparative and thematic analyses are often used in the same project, with the researcher moving backwards and forwards between transcripts, memos, notes and the research literature. This process is illustrated in Example 9.

EXAMPLE 9: RICHARD

Once Richard had discovered that members of the public who had close contact with higher educational institutions had clearer perceptions than those who had no contact, he felt two issues were important. First, he wanted to find out how close the contact had to be for people to have very clear perceptions of university, and second, he wanted to find out where perceptions came from for those people who had no contact with higher education.

Through careful choice of interviewee, and through comparing and contrasting the data from each transcript, he was able to develop a sliding scale of contact with higher education. This ranged from no contact, ever, for any member of the family or friends, through to personal contact by the interviewee attending higher education.

Having placed each interviewee somewhere on the scale, he then went back to the transcripts to look for hints about how their perceptions had been formed. At the same time he consulted existing research literature that addressed the issue of influences on personal perception to see if this would give him further insight into what was arising from his data.

After this process, if data was missing or he was unable to understand something that had been said, he would conduct another interview until he felt that his analysis, and his understanding, were complete.

Content analysis

For those types of analyses at the other end of the qualitative data continuum, the process is much more mechanical, with the analysis being left until the data has been collected.

Perhaps the most common method of doing this is to code by content. This is called **content analysis**. Using this method the researcher systematically works through each transcript assigning codes, which may be numbers or words, to specific characteristics within the text. The researcher may already have a list of categories or she may read through each transcript and let the categories emerge from the data. Some researchers may adopt both approaches, as Example 10 illustrates.

This type of analysis can be used for open-ended questions that have been added to questionnaires in large quantitative surveys, thus enabling the researcher to quantify the answers.

EXAMPLE 10: TINA

In her research on students' attitudes towards alcohol, Tina, from her own experience, felt that money, social life, halls of residence and campus bars would all be significant. She assigned code numbers to these issues and then went through each transcript, writing the code number above the relevant section when any of these issues were mentioned. Sure enough, they did appear to be important and were discussed in every interview, even with non-students.

However, she also found that many other issues were being discussed that she had not thought about previously, such as peer pressure and distance from home. As each new issue was mentioned, she ascribed another code and went back to previous transcripts to see if it had arisen but had been missed during the initial analysis. Although she had to return to the transcripts many times, this meant that by the end of the process Tina had completed a thorough analysis of her data.

Discourse analysis

Falling in the middle of the qualitative analysis continuum is **discourse analysis**, which some researchers have named **conversational analysis**, although others would argue that the two are quite different.

These methods look at patterns of speech, such as how people talk about a particular subject, what metaphors they use, how they take turns in conversation, and so on. These analysts see speech as a performance; it performs an action rather than describes a specific state of affairs or specific state of mind.

Much of this analysis is intuitive and reflective, but it may also involve some form of counting, such as counting instances of turn-taking and their influence on the conversation and the way in which people speak to others.

EXAMPLE 11: JULIE

Julie wanted to find out about women's experiences of premenstrual tension (PMT). As PMT was, at the time, a relatively new phrase to describe this condition, Julie was interested in finding out how women spoke about the problems they were experiencing, both in the present day and in the past. She wanted to look closely at what women from different generations said about themselves and how they talked to each other about their problems.

She decided to conduct five interviews and one focus group, and then analyse them using discourse analysis, which meant that she would break down each transcript into tiny parts. In the interview transcripts she looked for cultural, social and historical clues. In the focus group transcript she was interested in looking at how the women took turns to talk about the subject, especially among the different age groups.

Julie thought about her own position as a female researcher and how this might affect both what was being said and her interpretation of the data. Her final report contained large amounts of transcript to illustrate the points she had raised.

The processes of qualitative data analysis

These examples show that there are different processes involved in qualitative data analysis.

- You need to think about the data from the moment you start to collect the information.
- You need to judge the value of your data, especially that which may come from dubious sources.
- As your research progresses you need to interpret the data so that you, and others, can gain an understanding of what is going on.
- Finally, you need to undertake the mechanical process of analysing the data.

It is possible to undertake the mechanical process using software, which can save you a lot of time, although it may stop you becoming really familiar with the data. There are many dedicated qualitative analysis programs of various kinds available to social researchers that can be used for a variety of different tasks. For example, software could locate particular words or phrases; make lists of words and put them into alphabetical order; insert key words or comments; count occurrences of words or phrases or attach numeric codes. Some software will retrieve text, some will analyse text and some will help to build theory. Although a computer can undertake these mechanical processes, it cannot think about, judge or interpret qualitative data (see Table 10).

Analysing quantitative data

If you have decided that a large survey is the most appropriate method to use for your research, by now you should have thought about how you're going to analyse your data. You will have checked that your questionnaire is properly constructed and worded, you will have made sure that there are no variations in the way the forms are administered, and you will have checked over and over again that there is no missing or ambiguous information. If you have a well-designed and well-executed survey, you will minimise problems during the analysis.

Computing software

If you have software available for you to use, you should find this the easiest and quickest way to analyse your data. However, data input can be a long and laborious process, especially for those who are slow on the keyboard, and, if any data is entered incorrectly, it will influence your results. Large-scale surveys conducted by research companies tend to use questionnaires that can be scanned, saving much time and money, and you should check whether this option is available.

If you are a student, spend some time getting to know what equipment is available for your use, as you could save yourself a lot of time and energy by adopting this approach. Also, many software packages produce professional graphs, tables and pie charts that can be used in your final report, again saving a lot of time and effort.

Most colleges and universities run statistics courses and data analysis courses. Or the computing department will provide information leaflets and training sessions on data analysis software. If you have chosen this route, try to get onto one of these courses, especially those which have a 'hands-on' approach, as you might be able to analyse your data as part of your course work. This will enable you to acquire new skills and complete your research at the same time.

Statistical techniques

For those who do not have access to data analysis software, a basic knowledge of statistical techniques is needed to analyse your data. If your goal is to describe what you have found, all you need to do is count your responses and reproduce them. This is called a **frequency count** or **univariate analysis**. Table 11 shows a frequency count of age.

From this table you would be able to see clearly that the 20–29 age group was most highly represented in your survey. This type of frequency count is usually the first step in any analysis of a large-scale survey, and forms the base for many other statistical techniques that you might decide to conduct on your data (see Example 12).

TABLE 11: USING COMPUTERS FOR QUALITATIVE DATA ANALYSIS: ADVANTAGES AND DISADVANTAGES

ADVANTAGES	DISADVANTAGES
Using computers helps to alleviate time-consuming and monotonous tasks of cutting, pasting and retrieval of field notes and/or interview transcripts.	In focus groups the group moves through a different sequence of events that is important in the analysis but that cannot be recognised by a computer.
Computers are a useful aid to those who have to work to tight deadlines.	Programs cannot understand the meaning of text.
Programs can cope with both multiple codes and overlapping codes that would be very difficult for the researcher to cope with without the aid of a computer.	Software can only support the intellectual processes of the researcher – they cannot be a substitute for these processes.
Some software can conduct multiple searches in which more than one code is searched much more quickly and efficiently than by the researcher.	Participants can change their opinions and contradict themselves during an interview. A computer will not recognise this.
Programs can combine codes in complex searches.	The software might be beyond an individual's budget.
Programs can pick out instances of pre-defined categories that have been missed by the researcher during the initial analysis.	User-error can lead to undetected mistakes or misleading results.
Computers can be used to help the researcher overcome 'analysis block'.	Using computers can lead to an over-emphasis on mechanical procedures.

TABLE 12: AGE OF RESPONDENTS

AGE GROUP	FREQUENCY
Under 20	345
20–29	621
30–39	212
40–49	198
50–59	154
Over 59	121

However, there is a problem with missing answers in this type of count. For example, someone might be unwilling to let a researcher know their age, or someone else could have accidentally missed out a question. If there are any missing answers, a separate 'no answer' category needs to be included in any frequency count table. In the final report, some researchers overcome this problem by converting frequency counts to percentages that are calculated after excluding missing data. However, percentages can be misleading if the total number of respondents is fewer than 40.

EXAMPLE 12: TOM

Tom works part-time for a charity that provides information and services for blind and partially sighted people in the town. He was asked to find out how many people use the service and provide a few details about who these people are and what they do in life. Tom designed a short questionnaire that could be administered face-to-face and over the telephone by the receptionist. Anyone who called in person or telephoned the centre over a period of a month was asked these questions. If they had already completed a questionnaire they did not have to do so again.

Tom did not have access to any computing facilities, so he decided to analyse the questionnaires by hand. He conducted a count of gender, age, occupation, postcode area of residence and reason for attending or telephoning the centre. From this information, members of staff at the centre were able to find out that their main customers were women over the age of retirement. This meant that they were able to arrange more activities that suited this gender and age group. Tom found out also that one of the main reasons for contacting the centre was for more information on disability benefits. A Braille booklet and a digital recording containing all the relevant information was produced and advertised locally.

It took Tom one month to design and pilot the questionnaire, another month to administer the questionnaire, and two months to analyse the results and write the report.

Finding a connection

Although frequency counts are a useful starting point in quantitative data analysis, you may find that you need to do more than merely describe your findings. Often you will need to find out if there is a connection between one variable and a number of other variables. For example, a researcher might want to find out whether there is a connection between watching violent films and aggressive behaviour. This is called **bivariate analysis**.

In **multivariate analysis** the researcher is interested in exploring the connections among more than two variables. For example, a researcher might be interested in finding out whether women aged 40–50, in professional occupations, are more likely to try complementary therapies than younger, non-professional women and men from all categories.

Measuring data

Nominal scales

To move beyond frequency counts, it is important to understand how data is measured. In **nominal scales** the respondent answers a question in one particular way, choosing from a number of mutually exclusive answers. Answers to questions about marital status, religious affiliation and gender are examples of nominal scales of measurement. The categories include everyone in the sample, no one should fit into more than one category and the implication is that no one category is better than another.

Ordinal scales

Some questions offer a choice but from the categories given it is obvious that the answers form a scale. They can be placed on a continuum, with the implication being that some categories are better than others. These are called **ordinal scales**. The occupationally based social scale which runs from 'professional' to 'unskilled manual' is a good example of this type of scale. In this type of scale it is not possible to measure the difference between the specific categories.

Interval scales

Interval scales, on the other hand, come in the form of numbers with precisely defined intervals. Examples included in this type of scale are the answers from questions about age, number of children and household income. Precise comparisons can be made between these scales.

Arithmetic mean

In mathematics, if you want to find a simple average of the data, you would add up the values and divide by the number of items. This is called an **arithmetic mean**. This is a straightforward calculation used with interval scales where specific figures can be added together and then divided.

TIP

Be careful when making conclusions about cause and effect – that is, where you conclude that one thing is causing another. To check whether this really is the case you may need to address the following questions.

- When one variable (the presumed cause) changes, does the other variable (the presumed effect) also change?
- Are there other factors (or variables) that could have an effect?
- Does the presumed cause occur just before, or at a similar time to, the presumed effect?
- If you repeat your research, are the results the same?

However, it is possible to mislead with averages, especially when the range of the values may be great. Researchers, therefore, also describe the **mode**, which is the most frequently occurring value, and the **median**, which is the middle value of the range. The mode is used when dealing with nominal scales, for example it can show that most respondents in your survey are Catholics. The median is used when dealing with both ordinal and interval scales.

Quantitative data analysis can involve many complex statistical techniques that cannot be covered in this book. If you wish to follow this route you should read some of the data analysis books recommended below.

Summary

- The methods you use to analyse your data will depend upon whether you have chosen to conduct qualitative or quantitative research.

- For quantitative data analysis, issues of validity and reliability are important.

- Qualitative data analysis is a very personal process. Ask two researchers to analyse a transcript and they will probably come up with very different results.

- After having conducted an interview or a focus group, it is useful to complete a summary form that contains details about the interview. This can be attached to the transcript and can be used to help the analysis.

- Qualitative data analysis methods can be viewed as forming a continuum from highly qualitative methods to almost quantitative methods, which involve an element of counting.

- Examples of qualitative data analysis include thematic analysis, comparative analysis, discourse analysis and content analysis.

- The analysis of large-scale surveys is best done with the use of statistical software, although simple frequency counts can be undertaken manually.

- Data can be measured using nominal scales, ordinal scales or interval scales.

- A simple average is called an arithmetic mean; the middle value of a range is called the median; the most frequently occurring value is called the mode.

Further reading

Qualitative analysis

Bazeley, P. (2013) *Qualitative Data Analysis: Practical Strategies*, London: Sage.
Miles, M., Huberman, M. and Saldana, J. (2014) *Qualitative Data Analysis: A Methods Sourcebook*, 3rd edition, Thousand Oaks, CA: Sage Publications, Inc.
Silver, C. and Lewins, A. (2014) *Using Software in Qualitative Research: A Step-By-Step Guide*, 2nd edition, London: Sage.
Silverman, D. (2015) *Interpreting Qualitative Data*, 5th edition, London: Sage.

Quantitative analysis

Cramer, D. (2003) *Advanced Quantitative Data Analysis*, Maidenhead: Open University Press.
Marsh, C. and Elliot, J. (2008) *Exploring Data: An Introduction to Data Analysis for Social Scientists*, 2nd edition, Cambridge: Polity Press.
Sawilowsky, S. (ed) (2007) *Real Data Analysis*, Charlotte, NC: Information Age Publishing.
Treiman, D. (2009) *Quantitative Data Analysis: Doing Social Research to Test Ideas*, San Francisco, CA: John Wiley and Sons, Inc.

13
How to Report Your Findings

Once you have completed your research and analysed your data, there are three main ways of reporting your findings: written reports and journal articles, both of which can be reproduced online, and oral presentations.

Writing reports

If you are a student, your college or university may have strict rules and guidelines that you have to follow when writing up your report. You should find out what these are before you start your research, as this could influence your research methodology, as Jeanne found out (see Example 13).

EXAMPLE 13: JEANNE

I am a mature student and had worked for many years in a women's refuge prior to taking up my course. Naturally when it came to doing my dissertation I wanted to do some research within the refuge. I was interested in issues of women helping themselves to run the refuge rather than having inappropriate activities imposed upon them, sometimes by social workers who really had no experience of what the women were going through. That's when I found out about action research. I decided I would be able to work with the women to achieve acceptable goals for everyone.

In my opinion the research went really well. During the evaluation stage all the women said they were happy with both the process and the outcome. We were all happy and I was pleased with what we'd achieved. Then it came to writing my dissertation.

I had known all along that writing up a piece of action research would be

> difficult, but I had got my head around it and worked out how it could be done. Then I found out that my university had set rules for the format of a dissertation, and worse, my tutor had not even mentioned this when I started my research. So, I had to try and fit my research into what I saw as a really old-fashioned, scientific format which really didn't suit my work. I felt this was unfair and wouldn't do justice to the research I had actually carried out. I felt that I would have to spend so long justifying my methodology that there wouldn't be any room for anything else.
>
> At the moment, I've decided to argue my case at the examination committee and the Students' Union has agreed to represent me. I can't help feeling this will prejudice people against me. It has made me wonder what research is for and who it should benefit.

As pressures of work increase, tutors may not have the time to impart all the required information to each individual student. As a student you need to make sure that you have all the relevant information to hand. If you have not been given a copy of the dissertation guidelines, ask your tutor if they are available and from where they can be obtained. It is then up to you whether you want to follow these guidelines and conduct a piece of research that will fit well into the set format, or whether you have a burning passion to conduct something a little more innovative and become a trailblazer in the process. If the latter appeals to you, always talk over your ideas with your tutor first, as you could waste time and effort in conducting a piece of research that will not be considered suitable by the examiners.

If you are not a student, you may have more flexibility in the style and structure of your report. However, remember that one of the purposes of your report is to convince people that you have produced a good, sound piece of research. The more professional your report looks the better your chances of success, especially if you hope to aid decision making.

Remember the audience

An important point to remember when writing a report is to think about your audience. When doing this you may find it useful to ask the following questions:

- What style would your audience prefer?

- Are they likely to understand complex statistics or do you need to keep it simple?
- Have they the time to read through reams of quotations or are they interested only in conclusions and recommendations?
- Are they interested in your methodology? Do you need to justify your methodology to a non-believer or sceptic?
- Do you need to write using complex terminology or do you need to keep your language as simple as possible? (Normally I would recommend using plain, clear language, but on some occasions you will need to convince people of your knowledge of the subject by including some more complex terms. However, make sure you understand terminology thoroughly yourself. A few researchers have come unstuck by including terms that it becomes obvious later they do not understand.)

Structuring reports

Traditional written reports tend to be produced in the following format.

Title Page

This contains the title of the report, the name of the researcher and the date of publication. If the report is a dissertation or thesis, the title page will include details about the purpose of the report, for example 'A thesis submitted in partial fulfilment of the requirements of Sheffield Hallam University for the degree of Doctor of Philosophy'. If the research has been funded by a particular organisation, details of this may be included on the title page.

Contents Page

In this section the contents of the report is listed either in chapter or section headings with sub-headings, if relevant, and their page numbers.

List of Illustrations

This section includes title and page number of all graphs, tables, illustrations, charts, etc.

Acknowledgements

You may wish to acknowledge the help of your research participants, tutors, employers and/or funding body.

Abstract/Summary

This tends to be a one-page summary of the research, its purpose, methods, main findings and conclusion.

Introduction

This section introduces the research, setting out the aims and objectives, terms and definitions. It includes a rationale for the research and a summary of the report structure.

Background

This section includes all your background research, which may be obtained from the literature, from personal experience or both. You must indicate from where all the information to which you refer has come, so remember to keep a complete record of everything you read.

If you do not do this, you could be accused of plagiarism, which is a form of intellectual theft. When you are referring to a particular book or journal article, find out the accepted standard for referencing from your institution (see below).

Methodology and Methods

This section includes a description of, and justification for, the chosen methodology and research methods. The length and depth of this section will depend upon whether you are a student or employee. If you are an undergraduate student, you will need to raise some of the methodological and theoretical issues pertinent to your work, but if you are a postgraduate student you will need also to be aware of the epistemological and ontological issues involved.

If you are an employee, you may only need to provide a description of the methods you used for your research, in which case this section can be titled 'Research Methods'. Remember to include all the practical information people will need to evaluate your work, for example how many people took part, how they were chosen, your time scale, and data recording and analysis methods.

Findings/Analysis

This section includes your main findings. The content of this section will depend on your chosen methodology and methods. If you have conducted a large quantitative survey, this section may contain tables, graphs, pie charts and associated statistics. If you have conducted a qualitative piece of research this section may consist of descriptive prose containing lengthy quotations.

Conclusion

In this section you sum up your findings and draw conclusions from them, perhaps in relation to other research or literature.

Recommendations

Some academic reports will not need this section. However, if you are an employee who has conducted a piece of research for your company, this section could be the most important part of the report. It is for this reason that some written reports contain the recommendations section at the beginning of the report. This section lists clear recommendations that have been developed from your research.

Further Research

It is useful in both academic reports and work-related reports to include a section that shows how the research can be continued. Perhaps some results are inconclusive, or perhaps the research has thrown up many more research questions that need to be addressed. It is useful to include this section because it shows that you are aware of the wider picture and that you are not trying to cover up something that you feel may be lacking from your own work.

References

Small research projects will need only a reference section. This includes all the literature to which you have referred in your report. Find out which referencing system your college or university uses. A popular method is the Harvard system, which lists the authors' surnames alphabetically, followed by their initials, date of publication, title of book in italics, place of publication and publisher.

If the reference is a journal article, the title of the article appears in inverted commas and the name of the journal appears in italics, followed by the volume number and pages of the article. This is the method used in this book. Figure 5 provides a section of a bibliography from a PhD thesis to illustrate this method.

At this present time the way you are asked to reference material from the internet varies, so speak to your tutor about what information is required. In general, this should include the author's name, the date the work was created, the title of the page and/or the title of the work, the URL and the data you accessed the site.

Bibliography

Larger dissertations or theses will require both a reference section and a bibliography. As discussed above, the reference section will include all those publications to which you have referred in your report. If, however, you have read other work in relation to your research but not actually referred to them when writing up your report, you might wish to include them in a bibliography. However, make sure they are still relevant to your work – including books to make your bibliography look longer and more impressive is a tactic that won't impress examiners.

Clegg, S. (1985) 'Feminist Methodology: Fact or Fiction?' *Quality and Quantity*, 19: 83–97.
Cohen, A.P. (1994) *Self Consciousness: An Alternative Anthropology of Identity*, London: Routledge.
Cook, J.A. and Fonow, M.M. (1986) 'Knowledge and Women's Interests: Issues of Epistemology and Methodology in Feminist Sociological Research', *Sociological Enquiry*, 56: 2–29.
Erikson, E.H. (ed) (1978) *Adulthood*, New York: Norton.
Evans, N. (ed) (1980) *Education Beyond School: Higher Education for a Changing Context*, London: Grant McIntyre.
Faludi, S. (1992) *Backlash: The Undeclared War Against Women*, London: Chatto and Windus.

Fig. 5. Example list of references

Appendices

If you have constructed a questionnaire for your research, or produced an interview schedule or a code of ethics, it may be useful to include them in your report as an appendix.

In general, appendices do not count towards your total amount of words, so it is a useful way of including material without taking up space that can be used for other information. However, do not try filling up your report with irrelevant appendices, as this will not impress examiners. When including material, you must make sure that it is relevant – ask yourself whether the examiner will gain a deeper understanding of your work by reading the appendix. If not, leave it out.

Other information which could be included as an appendix are recruitment leaflets or letters; practical details about each research participant; sample transcripts (if permission has been sought); list of interview dates; relevant tables and graphs or charts that are too bulky for the main report.

TIP

If you suffer from writer's block, the following strategies might help.

- Take a break, get some fresh air and return to your writing when you feel refreshed.

- Brainstorm topics for each section of your report and write them into paragraphs at a later stage.

- Put together parts of your report that don't require thoughts and ideas, such as your reference section or the bibliography.

- Write a draft, however bad. You can return to your draft at a later stage when you feel more creative.

- Don't be afraid to write your report in the wrong order. For example, often the introduction is the hardest part and can prevent you from starting your report. Try leaving the introduction until later and get on with other sections that you might find easier.

TEN REASONS WHY REPORTS FAIL

- There is no logical structure
- Ideas are not well thought out
- Work is disorganised
- Assumptions are made that cannot be justified by evidence
- There are too many grammatical and spelling mistakes
- Sentences and/or paragraphs are too long or too obscure
- It is obvious that ideas and sentences have been taken from other sources
- There is too much repetition
- There is too much irrelevant information
- Summary and conclusions are weak

Writing journal articles

If you want your research findings to reach a wider audience, it might be worth considering producing an article for a journal. Most academic journals do not pay for articles they publish, but many professional or trade publications do pay for your contribution, if published. However, competition can be fierce and your article will have to stand out from the crowd if you want to be successful. The following steps will help you to do this:

- Choose a topical, original piece of research.
- Do your market research – find out which journal publishes articles in your subject area.

- Check on submission guidelines – produce an article in the correct style and format and of the right length.

- Read several copies of the journal to get an idea about the preferences of editors.

- If you are thinking about writing for a trade publication, approach the editors by letter, asking if they might be interested in an article. Include a short summary of your proposed article.

- Produce a succinct, clear, interesting and well-written article – ask friends, tutors or colleagues to read it and provide comments.

- Make sure there are no mistakes, remembering to check the bibliography.

- If it is your first article, gain advice from someone who has had work published. Also you might find it easier to write an article with someone else – some tutors or supervisors will be willing to do this as it helps their publication record if their name appears on another article. You may find that you will do most of the work, but it is very useful to have someone read your article and change sections that do not work or read well. It is also useful to have people comment on your methodology or analysis assumptions that could be criticised by other researchers.

TIP

Beware of 'predatory' journals. These are open-access journals with questionable peer-review practices and aggressive marketing campaigns. They lack checks on quality and rigour, publishing papers rapidly for a high fee. Ensure that you assess the reputation and quality of journals before you write and submit a paper.

Producing oral presentations

Another method of presenting your research findings is through an oral presentation. This may be at a university or college to other students or tutors, at a conference to other researchers or work colleagues, or in a workplace to colleagues, employers or funding bodies. Many researchers find that it is better to provide both a written report and an oral presentation, as this is the most

TABLE 13: MAKING PRESENTATIONS: DOS AND DON'TS

DO	DON'T
Arrive early and make sure the room is set out in the way that you want. Make sure that all the equipment is available and that you know how to work it.	Rush in late, find that the overhead projector doesn't work and that you have no pen for the whiteboard.
Try to relax and breathe deeply. Acknowledge that this is your first presentation and people will tend to help you along.	Worry about showing your nerves. Everybody gets nervous when they first start giving presentations and your audience should know this.
Produce aide-mémoirs, either on cards, paper, OHP transparencies or presentation software such as PowerPoint.	Read straight from a paper you have written.
Make it clear from the outset whether you are happy to be interrupted or whether questions should be left for the end. If you have invited questions, make sure you make every effort to answer them.	Get cross if you are interrupted and have not mentioned that you don't want this to happen. Don't invite questions and then fail to answer them.
Look around the room while you are speaking – if it's a small group, make eye-contact with as many people as possible.	Look at your notes, never raising your head.
Present interesting visual information such as graphs, charts and tables in a format that can be viewed by everyone. This could be OHP transparencies, slides, PowerPoint or handouts.	Produce visual information that people can't see, either due to its size or print quality.
Alter the tone and pitch of your voice, length of sentence and facial/hand gestures to maintain audience interest. Show that you are interested in your subject.	Present in a monotone voice with no facial/hand gestures. Don't make it seem that your subject bores the pants off you.
Produce a paper or handout that people can take away with them.	Let the audience go home without any record of what you have said.
Talk to people after your presentation and ask them how it went, whether there are any improvements they might suggest for future presentations.	Run away never to be seen again.

effective way of enabling a wider audience to find out about the research, especially if you also reproduce your written report online.

If you want people to take notice of your results, you need to produce a good presentation. Table 13 provides a list of dos and don'ts when making a presentation.

TIP

When presenting at conferences or to work colleagues, try to anticipate questions and rehearse answers. It is useful to anticipate criticism so that you can practise arguments and prepare a defence of your research.

PowerPoint is a useful presentation graphics program, which enables you to create slides that can be shared live or online. You can enhance your presentation with animation, artwork and diagrams that make it more interesting for your audience. Full details about PowerPoint can be found at *www.office.microsoft.com*.

Summary

- There are three main ways of reporting your findings: written reports, journal articles and oral presentations.
- Before starting your research, find out whether you are going to be restricted by structure, style and content of your final report.
- Think about your audience and produce your report accordingly.
- A traditional written report includes the following:
  - title page
  - contents page
  - list of illustrations
  - acknowledgements
  - abstract/summary
  - introduction
  - background
  - methodology/methods
  - findings/analysis

- conclusions
- recommendations
- further research
- references
- bibliography
- appendices.

- If you are interested in writing an article for a journal, do your market research. Make sure that the subject matter, style, structure and length of your article suit the journal.

- Try to seek advice and comments from people experienced in writing journal articles.

- Think about producing your first article with another, more experienced researcher.

- When making oral presentations, always be prepared. Arrive early, make sure equipment works and that you have everything you need.

- Show that you are interested in what you are saying and try to keep audience interest by using visual aids and altering tone, pitch and gestures.

- Don't shrug off questions or patronise your audience – pitch your presentation at the right level.

- Never let an audience leave without taking away a record of what you have said.

Further reading

Bowden, J. (2011) *Writing a Report*, 9th edition, Oxford: How To Books.
Chivers, B. and Shoolbred, M. (2007) *A Student's Guide to Presentations*, London: Sage.
Gustavii, B. (2017) *How to Write and Illustrate a Scientific Paper*, 3rd edition, Cambridge: Cambridge University Press.
Smith, P. (2002) *Writing an Assignment*, 5th edition, Oxford: How To Books.

14
How to be an Ethical Researcher

As researchers we are unable to conduct our projects successfully if we do not receive the help of other people. If we expect them to give up their valuable time to help us, it follows that we should offer them something in return. Also, many people are willing to disclose a lot of personal information during our research, so we need to make sure that we treat both the participants and the information they provide with honesty and respect. This is called research ethics.

Treating participants with respect

As a researcher you must remember that the research process intrudes on people's lives. Some of the people who take part in your research may be vulnerable because of their age, social status or position of powerlessness. If participants are young, you need to make sure a parent or guardian is present. If participants are ill or reaching old age, you might need to use a proxy and care should be taken to make sure that you do not affect the relationship between the proxy and the participant.

Some people may find participation a rewarding process, whereas others will not. Your research should not give rise to false hopes or cause unnecessary anxiety. You must try to minimise the disruption to people's lives and if someone has found it an upsetting experience, you should find out why and try to ensure that the same situation does not occur again.

As a researcher you will encounter awkward situations, but good preparation and self-awareness will help to reduce these. If they do happen, you should not dwell too long on the negative side – reflect, analyse, learn by your mistakes and move on.

Anonymity and confidentiality

You must do your best to ensure anonymity and confidentiality. However, information given by research participants in confidence does not enjoy legal privilege. This means that the information may be liable to subpoena by a court. If you're dealing with very sensitive information that you know could be called upon by a court of law, you will need to inform your participants that you would be obliged to hand over the information.

Recognising overt and covert research

Overt research means that it is open, out in the public and that everyone knows who you are and what you are doing. Covert research means that you are doing it under cover, that no one knows you are a researcher or what you are doing. In my opinion covert research should be kept to a minimum – there are enough journalists and television personalities doing this kind of undercover, sensationalist work.

Covert research

In the past, researchers have justified their covert work by saying that it has been the only way to find out what goes on in a particular organisation that would not otherwise let a researcher enter. Such work has been carried out within religious cults and warring gangs of young people. However, this type of research can have serious implications for the personal safety of the researcher and the people with whom she comes into contact. It can also give research a bad name – other people may read about the work and become suspicious about taking part in future projects.

Overt research

I believe researchers should be open and honest about who they are and what they're doing. People can then make an informed choice about whether they take part in a project. It is their prerogative to refuse – nobody should be forced, bullied or cajoled into doing something they don't want to do.

If people are forced to take part in a research project, perhaps by their boss or someone else in a position of authority, you will soon find out. They will not be willing to participate and may cause problems for you by offering false or useless information, or by disrupting the data collection process. Who can blame them? Wouldn't you do the same if you were forced to do something you didn't want to do?

This means that not only you should be open and honest about who you are and what you're doing, but so should those who open the gates for you, especially those who are in a position of authority. Consider Example 14 from a student new to research.

> **TIP**
>
> 'Informed consent' in research is an agreement that is made by a participant to take part in your research, based on an understanding of what is involved. For this consent to be valid it must be informed, voluntary, free and given by a person who has the capacity to make the decision. A person must understand the purpose, benefits and potential risks of taking part in your research and must be given time to make their decision.

EXAMPLE 14: STEVE

It was the first project I'd ever done. I wanted to find out about a new workers' education scheme in a car factory. One of my tutors knew someone in charge of the scheme and that person arranged for me to hold a focus group in the factory. This meant that the person in charge of the scheme chose the people for the focus group. I was really pleased because it meant I didn't have to do a lot of work getting people to come. Of course I soon found out that he'd chosen these people for a particular reason, and he'd actually told them that they had to attend, that there was no choice involved.

When I turned up to hold the group, no one had been told who I was and what they were doing there. When I started to introduce myself, some of the workers looked a bit uneasy and others just looked plain defiant.

It was only after the group that I spoke to someone who said that they'd all thought I was a 'spy' for the company and that some had decided to give the 'company line' on what the scheme was all about, whereas others had decided not to say anything. She said that really they didn't believe a lot of what had been said, but none of them dared say anything different as they thought I was going to go straight to management with the results.

I felt that the information I collected wasn't very useful in terms of my research, but it was useful in terms of getting an idea about employer – employee relations.

Introducing yourself

If you are relying on someone else to find participants for you, it is important that you make sure that that person knows who you are and what you're doing, and that this information is then passed on to everyone else. A useful way to do this is to produce a leaflet that can be given to anyone who might be thinking about taking part in your research. This leaflet should contain the following information:

- Details of who you are (student and course or employee and position).
- Details of the organisation for which you work or at which you study.
- Information about who has commissioned/funded the research, if relevant.
- Information about your project – subject and purpose.
- Details about what will happen to the results.
- Information about the personal benefits to be gained by taking part in the project. This section is optional, but I find it helps to show that people will gain personally in some way by taking part in the research. This acts as an incentive. You might offer further information about something in which they are interested, or you might offer them a copy of the final report. Some consumer research companies offer entry into a prize draw or vouchers for local shops and restaurants.

Producing a Code of Ethics

Once you have been open and honest about what you are doing and people have agreed to take part in the research, it is useful to provide them with a Code of Ethics. The best time to do this is just before they take part in a focus group, or interview or experiment, or just before they fill in your questionnaire. The Code of Ethics supplies them with details about what you intend to do with the information they give and it shows that you intend to treat both them and the information with respect and honesty. It covers the following issues:

- **Anonymity**: you need to show that you are taking steps to ensure that what participants have said cannot be traced back to them when the final report is produced. How are you going to categorise and store the information? How are you going to make sure it is not easily accessible to anyone with unscrupulous intentions? Do you intend to change the names of people, towns and organisations? If not, how will you ensure that what someone says cannot be used against them in the future? However, you must be careful not to make promises that you cannot keep.

- **Confidentiality**: you need to show that information supplied to you in confidence will not be disclosed directly to third parties. If the information is supplied in a group setting, issues of confidentiality should be relevant to the whole group, who should also agree not to disclose information directly to third parties. You need to think about how you're going to categorise and store the information so that it cannot fall into unscrupulous hands. Again, you need to make sure that you do not make promises that you can't keep.

- **Right to comment**: this will depend on your personal methodological preferences and beliefs. Some researchers believe that willing participants should be consulted throughout the research process and that if someone is unhappy with the emerging results and report, they have the right to comment and discuss alterations. Indeed, this can be seen as part of the research process itself. Other researchers believe that once the information has been supplied, it is up to them what they do with it. If you're not willing to discuss the final report or take on board comments from unhappy participants, you must make this clear from the outset.

- **The final report**: it is useful for participants to know what is going to happen with the results. Who will receive a free copy of the report? Will it be on public display? If the final report is very long you can produce a shorter, more succinct report that can be sent to interested participants. This will keep down your own production and postage costs.

- **Data Protection**: you need to show that you understand the Data Protection

Act and that you intend to comply with its rules. The Data Protection Act 2018 came into force on 23 May 2018. It contains equivalent regulations and protections to the GDPR, which sets rules for processing personally identifiable information of individuals inside the European Union. It applies to all organisations and enterprises that undertake activities with the European Economic Area. More information about the Data Protection Act 2018 can be obtained from *ico.org.uk/for-organisations/data-protection-act-2018* and more information about the GDPR can be obtained from *ico.org.uk/for-organisations/guide-to-the-general-data-protection-regulation-gdpr*.

Personal data covers both facts and opinions about an individual. More details about the Data Protection Act can be found at *www.ico.org.uk* or from the address at the end of the book.

The amount of detail you provide in your Code of Ethics will depend on your research, your participants and your methodological preferences. Some people will not want to see a lengthy list of ethical considerations, whereas others will go through your list with a fine-tooth comb. It is for this reason that you might find it useful to produce two – a short summary and a longer version for those who are interested. A short Code of Ethics is provided in Figure 6.

The British Sociological Association has produced a Statement of Ethical Practice, which can be viewed at *www.britsoc.co.uk* or ordered from the address at the end of this book. This statement covers issues such as professional integrity; relations with, and responsibilities towards, research participants; relations with, and responsibilities towards, sponsors and/or funders. It is a very detailed list and will help you to think about all the ethical issues that may arise during your research.

Summary

- Our research would not be possible without the help and cooperation of other people. If we expect people to continue helping us, we should treat them with honesty and respect.

- Disruption to a participant's life should be kept to a minimum.

Anonymity

I guarantee that I will not use any names and addresses in the final report, or store or categorise information using names and addresses. This will help to ensure that what you have said during the discussion will not be traced back to you by third parties.

Confidentiality

I guarantee that I will not disclose directly any information provided in this group to third parties, unless permission has been granted to do so. As some of the comments made in this group may be of a personal or private nature, other participants should respect the confidentiality of individuals and also not disclose information directly to third parties.

Your right to comment

I agree to keep you informed about the progress of the research. If at any stage you wish to comment on the emerging results or final report you may do so. I agree to listen to your comments and make relevant alterations, if appropriate.

The final report

This research is funded by [name of organisation or funding body]. A copy of the final report will be sent to this organisation, to the University library and to anyone who has taken part in the research who has requested a copy.

Data Protection

The researcher will comply with the Data Protection Act 2018.

Fig. 6. Code of ethics

- False hopes or expectations should not be raised.
- Confidential or anonymous data do not enjoy legal privilege.
- Overt research means that it is out in the open – everyone knows who the researcher is and what she is doing.
- Covert research means that it is under-cover work. Nobody knows who the

researcher is and what she is doing. This type of work can give research a bad name and has personal safety implications for the researcher and for the people with whom she comes into contact.

- It is an individual's prerogative to refuse to take part in research – nobody should be forced, bullied or cajoled into taking part.

- If someone is thinking about helping with your research, they should be given a leaflet that includes the following information:
  – Details about who you are and the organisation for which you work.
  – Details about your project, the funding body and what will happen to the results.
  – Information about possible benefits to be gained by taking part in the research (false promises should not be made).

- A short Code of Ethics should be given to everyone who takes part in the research. This should include the following issues:
  – anonymity
  – confidentiality
  – right to comment
  – the final report
  – Data Protection.

- A longer, more detailed Statement of Ethical Practice can be produced for anyone who requests a copy.

Further reading

Comstock, G. (2013) *Research Ethics: A Philosophical Guide to the Responsible Conduct of Research*, New York: Cambridge University Press.
Lee-Trewick, G. and Linkogle, S. (eds) (2000) *Danger in the Field: Risk and Ethics in Social Research*, London: Routledge.
Miller et al. (eds) (2012) *Ethics in Qualitative Research*, 2nd edition, London: Sage.
Oliver, P. (2010) *The Student's Guide to Research Ethics*, 2nd edition, Maidenhead: Open University Press.
Ransome, P. (2013) *Ethics and Values in Social Research*, Basingstoke: Palgrave Macmillan.

Useful Addresses

For information about the Data Protection Act and other information about personal rights, contact:

Information Commissioner's Office
Wycliffe House
Water Lane
Wilmslow
Cheshire
SK9 5AF
ICO Helpline: 0303 123 1113
Email: *use contact form*
Website: *www.ico.org.uk*

For advice about research methods and ethics, contact:

The British Sociological Association (BSA)
Bailey Suite, Palatine House
Belmont Business Park
Belmont
Durham
DH1 1TW
Tel: 0191 383 0839
Email: *enquiries@britscoc.co.uk*
Website: *www.britsoc.co.uk*

For information about market, social and opinion research, contact:

The Market Research Society
15 Northburgh Street
London
EC1V 0JR
Tel: 020 7490 4911
Email: *info@mrs.org.uk*
Website: *www.mrs.org.uk*

For information about the conduct, development and application of social research, contact:

The Social Research Association
c/o Nuffield Foundation
24 Bedford Square
London WC1B 3JS
Tel: 020 7998 0304
Email: *admin@the-sra.org.uk*
Website: *www.the-sra.org.uk*

If you are interested in finding out about postgraduate research funding and training, contact one of the following research councils:

Economic and Social Research Council (ESRC)
Polaris House
North Star Avenue
Swindon
SN2 1UJ
Tel: 01793 413000
Email: esrcenquiries@esrc.ac.uk
Website: www.esrc.ac.uk

Biotechology and Biological Sciences Research Council (BBSRC)
Polaris House
North Star Avenue
Swindon
SN2 1UH
Tel: 01793 413200
Email: *eligibility@bbsrc.ac.uk*
Website: *www.bbsrc.ac.uk*

Engineering and Physical Sciences Research Council (EPSRC)
Polaris House
North Star Avenue
Swindon
SN2 1ET
Tel: 01793 444000
Email: *infoline@epsrc.ac.uk*
Website: *www.epsrc.ac.uk*

Medical Research Council (MRC)
Polaris House
North Star Avenue
Swindon
SN2 1UH
Tel: 01793 413200
Email: *grants@headoffice.mrc.ac.uk*
Website: *www.mrc.ac.uk*

Science and Technology Facilities Council
Polaris House
North Star Avenue
Swindon
SN2 1SZ
Tel: 01793 442000
Email: *enquiries@stfc.ac.uk*
Website: *www.stfc.ac.uk*

Natural Environment Research Council (NERC)
Postgraduate Support Group Awards and Training
Polaris House
North Star Avenue
Swindon
SN2 1EU
Tel: 01793 411500
Email: *see website for staff contacts*
Website: *www.nerc.ac.uk*

The Arts and Humanities Research Council (AHRC)
Whitefriars
Lewins Mead
Bristol
BS1 2AE
Tel: 0117 987 6500
Email: *enquiries@ahrc.ac.uk*
Website: *www.ahrc.ac.uk*

For information about funding for social research projects, contact:

Joseph Rowntree Foundation
The Homestead
40 Water End
York
YO30 6WP
Tel: 01904 629 241
Email: *info@jrf.org.uk*
Website: *www.jrf.org.uk*

The British Library can be contacted at:

The British Library
St. Pancras
96 Euston Road
London
NW1 2DB
Tel: 0330 333 1144
Email: *Customer-Services@bl.uk*
Website: *www.bl.uk*

For information on copyright law, contact:

The Copyright Licensing Agency
Barnards Inn
86 Fetter Lane
London
EC4A 1EN
Tel: 020 7400 3100
Email: *cla@cla.co.uk*
Website: *www.cla.co.uk*

Index